Editor
Kim Fields

Editorial Project Manager
Mara Ellen Guckian

Editor-in-Chief
Sharon Coan, M.S. Ed.

Illustrators
Kevin Barnes
Renée Christine Yates

Cover Artist
Barb Lorseyedi

Art Manager
Kevin Barnes

Art Director
CJae Froshay

Imaging
James Edward Grace
Rosa C. See
Richard Easley

Product Manager
Phil Garcia

Publishers
Rachelle Cracchiolo, M.S. Ed.
Mary Dupuy Smith, M.S. Ed.

Full Color Literacy Math Skills
Pre K–1

Author
Amy Decastro, M.A.

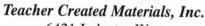

Teacher Created Materials, Inc.
6421 Industry Way
Westminster, CA 92683
www.teachercreated.com.
ISBN-0-7439-3398-2

©2004 Teacher Created Materials, Inc.

Made in U.S.A.

The classroom teacher may reproduce copies of materials in this book for classroom use only. The reproduction of any part for an entire school or school system is strictly prohibited. No part of this publication may be transmitted, stored, or recorded in any form without written permission from the publisher.

Table of Contents

Introduction 3

How to Use This Book 4

Standards and Benchmarks 6

Center Management 8

Problem Solving and Reasoning Centers

    A Class Act 9

    Dish It Out 17

Probability and Statistics Centers

    Barnyard Predictions 29

    The Survey Says 39

Measurement Centers

    Let's Go! 53

    Time for Fun 65

Patterns/Relations/Algebra and Functions Center

    Bugs! 79

Geometry Center

    Adding Shapes 93

Number Sense Centers

    Food for Thought 113

    Dino-Mite 129

Mathematical Connections Center

    Hop to It 143

Numeration and Estimation Center

    Penguin Pals 155

Introduction

Children begin building a basic understanding of math concepts at a very young age. Their environment is saturated with opportunities to use math. This easy-to-use math book was designed to capitalize on these everyday opportunities and provide meaningful, creative math experiences for young learners.

The lessons in the book were developed with a centers-approach to learning. Centers are an ideal way to expose young children to meaningful, enjoyable math activities. Young children learn from direct experiences where they have opportunities to explore. Most gain very little from a straight lecture format. They learn best in a hands-on environment that allows them to interact with materials, and with other children. Centers solidify skills introduced in whole-group time. Centers also encourage the development of critical-thinking and decision-making skills.

A classroom with an established center format gives teachers more time for one-on-one interaction with students, allows for different learning styles, and encourages positive interaction among students. Centers give children an opportunity to explore, discover, practice, apply skills, and become independent thinkers. Students are usually paired up or in small groups while working at centers. This collaboration helps children learn to articulate thoughts and share approaches to solving problems, as well as see solutions that may be quite different from their own.

Literacy Centers for Math Skills provides developmentally appropriate, hands-on activities that enrich a child's level of math comprehension while meeting current kindergarten benchmarks and standards.

How to Use This Book

The activities in *Literacy Centers for Math Skills* provide a fun way for students to master math standards. Lessons can be modified to meet the needs of the students involved. Prior knowledge and experiences will need to be taken into account. Materials needed to teach the lesson are listed along with a detailed lesson plan. Each center includes:

Teacher Page

The information included here will guide the teacher in implementing the center. It includes the lesson objective, the materials needed, ideas for the center setup, the lesson or presentation, and an extension(s). It can be laminated and placed in the center for adult helpers and teacher assistants to use as a reference. Explanations of each section of the teacher page are as follows:

- **Objective(s)**—The activities in this book are based on National Math Standards. The objective(s) corresponding to a particular standard(s) will be noted at the beginning of each activity.

- **Materials**—Materials specific to each center are listed. Many of the materials are ready to use within the book and may only need to be laminated and/or cut apart for students. While most other items can be found in the classroom, some may need to be ordered or collected ahead of time.

- **Presentation**—This section describes how the center will meet the cited standard(s). An explanation of the center activity is provided as well as ways in which to incorporate the featured vocabulary words. Descriptions and definitions are given when appropriate.

- **Extension(s)**—This section offers additional independent reinforcement of the standard taught within the previous lesson. It was designed for students who may finish early or need a little more practice to help master that particular standard.

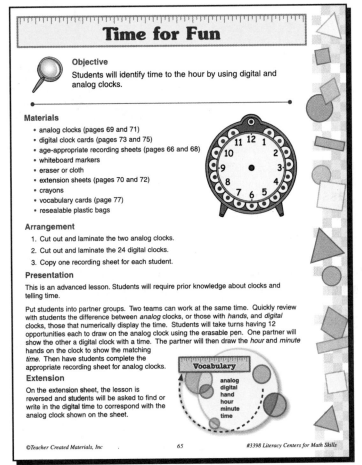

How to Use This Book

Activity Card

Each center will include a full-color, step-by-step, illustrated card to guide the student through the activity. The activity cards have directions similar to those used in the teacher presentation, but the text has been simplified. This card should be laminated. Place the card in the center for students to refer to when necessary.

Show the class how the illustrated activity cards guide them through the activity with step-by-step reminders. Ask for questions or clarifications. Discuss how to take turns and what to do while waiting.

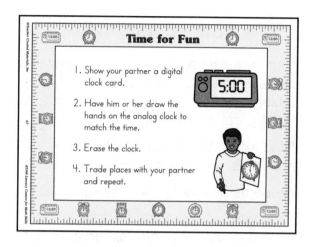

Assessment Pages

The reproducible student recording sheets were created to assist teachers in gathering information about what students have learned at the center. Choose the assessment pages (worksheets) that best suit your students needs and/or abilities. These pages can serve as assessment tools for the teacher in conjunction with classroom discussions. It is important for students to share and discuss their observations and findings. Reviewing each hands-on activity helps cement the learning and increases retention.

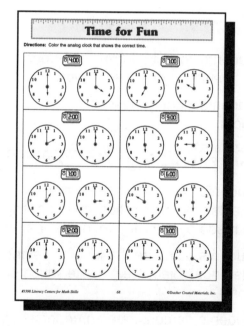

Vocabulary Cards

Developing appropriate mathematical vocabulary is an important outcome of these lessons. To help facilitate this, vocabulary cards are included for each center. Keep in mind that the purpose of the cards is to expose children to the language and the meaning of the words, not necessarily the spelling. If appropriate, the vocabulary cards could be placed in the writing and journaling areas or on an interactive Math Word Wall. This will encourage verbal vocabulary building, as well as written usage.

  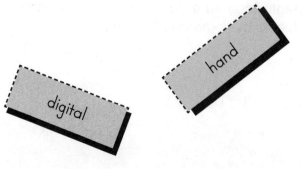

Standards and Benchmarks

The standards addressed in this book are derived from those recommended by the National Council of Teachers of Mathematics.

Problem Solving and Reasoning

- Students communicate their understanding of the problem. They choose an appropriate strategy to solve the problem, such as: use manipulatives or sketches; make a graph, chart or table; use an organized list; act it out; draw a picture; look for patterns; choose an operation; or use logical reasoning.
- Students make decisions about how to set up a problem. They determine the approach, materials, and/or strategies to be used.
- Students use tools and strategies, such as manipulatives or sketches, to model problems.
- Students solve problems and justify their solutions by explaining the reasoning used with concrete objects and/or pictorial representations. They make precise calculations and check the validity of the results in the context of the problem.

Probability and Statistics

- Students collect information about objects by posing questions, collecting data, and recording the results using objects, pictures, and picture graphs.
- Students identify, describe, and extend simple patterns by referring to their shapes, sizes, or colors.

Measurement and Geometry

- Students understand the concept of time and units used to measure it; they understand that objects have properties, such as length, weight, and capacity; and realize that comparisons may be made by referring to those properties.
- Students compare the length, weight, and capacity of objects by making direct comparisons with reference objects (e.g., note which object is shorter, longer, taller, lighter, heavier, or holds more).
- Students demonstrate an understanding of concepts of time and tools that measure time. They name the days of the week and identify the time (to the nearest hour) of everyday events.
- Students identify common objects in their environment and describe the geometric features by identifying and describing common geometric objects and comparing familiar plane and solid objects by common attributes.

Standards and Benchmarks

Patterns/Relations/Algebra and Functions

➥ Students sort and classify objects by attribute and identify objects that do not belong to a particular group.

➥ Students copy, extend, and design kinesthetic, visual, and whole number patterns and be able to recognize patterns in everyday life.

➥ Students count by 1's and 10's to 100, and by 5's to 50. Students identify numbers to 20 and symbols (+, −, =) in equations.

Number Sense

➥ Students understand the relationship between numbers and quantities. They compare two or more sets of objects up to 10 and identify which set is equal to, more than, or less than the other. They count, recognize, represent, name, and order a number of objects up to 30. They know that the larger numbers describe sets with more objects than the smaller number sets.

➥ Students understand and describe simple additions and subtractions.

➥ Students use concrete objects to determine the answers to addition and subtraction problems for two numbers that are each less than 10.

➥ Students use estimation strategies in computation and problem solving involving numbers that use the ones and tens places and recognize when an estimate is reasonable.

Mathematical Connections

➥ Students recognize and use equivalent representations of the same math concept (concrete, pictorial, abstract). They recognize the use of mathematical concepts, ideas, and patterns in other subjects.

➥ Students make connections with all math areas and concepts such as connecting patterning to skip counting. They apply mathematical thinking and language to real-world situations and relate mathematics to their daily life.

Numeration and Estimation

➥ Students model, identify, read, write, order, and compare whole numbers. They model and identify ordinal numbers 1st–10th.

➥ Students group objects into fives and tens, estimate quantities and find actual amounts, and count backward from 10.

➥ Students demonstrate one-to-one correspondence to 20.

➥ Students model and identify equal and unequal portions.

➥ Students identify ½ of regions and sets.

Center Management

Successful center management can dramatically affect students' attitudes and habits of learning. The center environment should be organized, stimulating, and comfortable. Creating such an environment entails arranging a practical physical layout, supplying diverse materials and supplies, and encouraging students to have a sense of belonging and ownership. Centers should provide opportunities for children to have meaningful educational experiences and be actively involved in their own learning.

When planning centers, it is important to consider how the centers will coincide with your own philosophy and comfort level. Begin the process slowly if need be. Once you gain confidence in the process, you will be able to increase the number of lessons and standards integrated into centers. There is no right or wrong way to approach centers. The main things to consider are: *What do you want to accomplish by using a centers approach to teaching in your classroom?* and *How will students benefit from this approach?*

Once you have answered those questions, you are ready to move on to efficiently setting up the room. Create a logical traffic flow by arranging tables, desktops, or floor space so that students are able to roam freely to their center and have easy access to supplies. You may have noisy, active centers in one area of the room and quieter, more structured centers in another area. The goal is to be as creative as you can to make it a comfortable environment for students.

The role of the teacher prior to center time is to present new centers and model appropriate procedures. Introduce any new tools or materials and demonstrate proper usage. Use the vocabulary cards provided in each section to explain the center. Laminate and post the student card at the center. During center times, a teacher can be actively involved in the students' learning process. Teachers can observe, monitor, and work in small groups on skill instruction, remedial skills, or enrichment activities. This is also a good time to ask questions and take anecdotal records.

The amount of daily class time spent at centers, again, depends on each teacher's comfort level and needs. The key is to establish blocks of time so students do not feel rushed or need to be interrupted. Time spent at centers can be 30 minutes a day or it may be used as an integral part of your program and can be used in 1–2 hour increments throughout the day.

Teachers may choose to have a few centers or more. Again, it all depends on your comfort level and what it is that you want to accomplish. For interaction and collaboration purposes, it is important to have at least two students at every center. That number can be higher and will depend on the number of children in your classroom and the activities that are available at one time.

A Class Act

Objective

Students will develop logical thinking by arranging pictures into groups of attributes. They will justify and explain the reasoning used with their self-portraits.

Materials

- class portrait template (page 10)
- attribute chart (page 13)
- age-appropriate recording sheets (pages 12 and 14)
- crayons
- vocabulary cards (page 15)

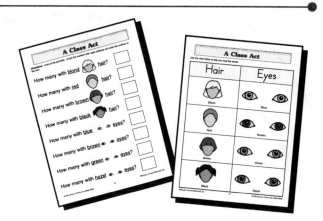

Arrangement

1. Copy one class portrait template and one recording sheet for each student.
2. Introduce the attribute chart to the class.
3. Allow time for the entire class to complete their features on the face template (eye color and hair color).

Presentation

This center encourages the understanding of sorting based on individual *attributes* (characteristics). Students will identify *different* attributes (such as eye color and hair color) of classmates by *comparing* each of the portraits (previously drawn by the students). They will sort one attribute at a time. Discuss the different color possibilities for hair (red, brown, black, and blond) and eyes (blue, hazel, green, and brown) and give examples.

Explain to students that they will *sort* the portraits into groups of the *same* attribute. Then they will *count* the number of students in each group by using one-to-one correspondence and number representation on the given recording sheet. Be sure to demonstrate how to record responses on the appropriate recording sheet.

Extension

Allow students to create a comparison pattern by having one cube symbolize an attribute from the self-portraits. For example, assign a blue cube for each blond student. For each student with brown eyes, assign a brown cube. Students connect the cubes; then count the number of each attribute.

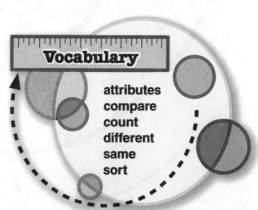

Vocabulary

attributes
compare
count
different
same
sort

A Class Act

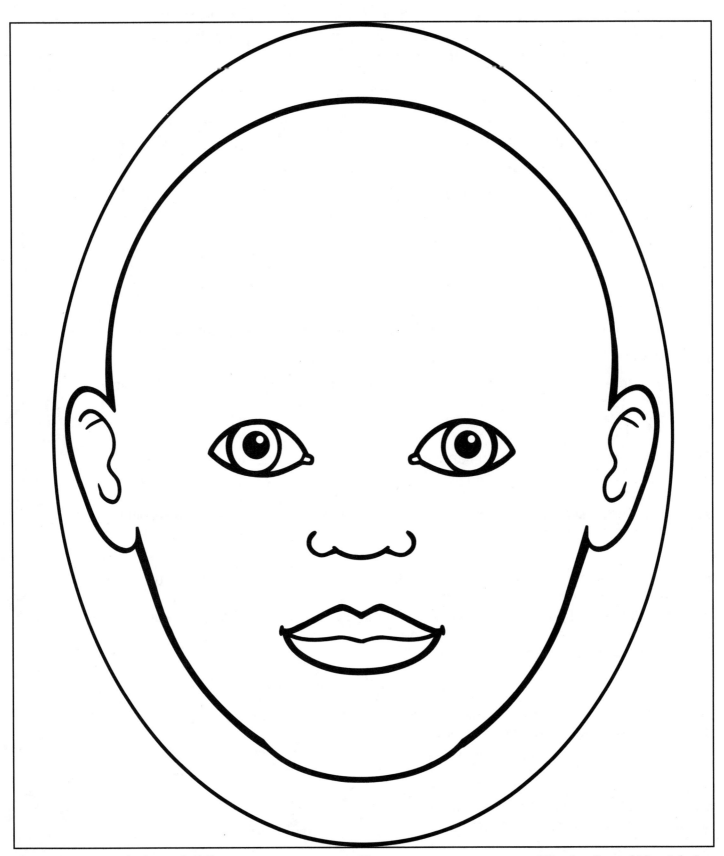

A Class Act

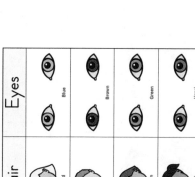

1. Sort the portraits according to the different attributes.

2. Answer the questions on the recording sheet.

3. Use the Hair and Eyes chart to help you read the underlined words.

A Class Act

Directions: Color in one box for each classmate who has that attribute.

- blond
- red
- brown
- black
- blue
- brown
- green
- hazel

A Class Act

Directions: Use the chart below to help you read the words.

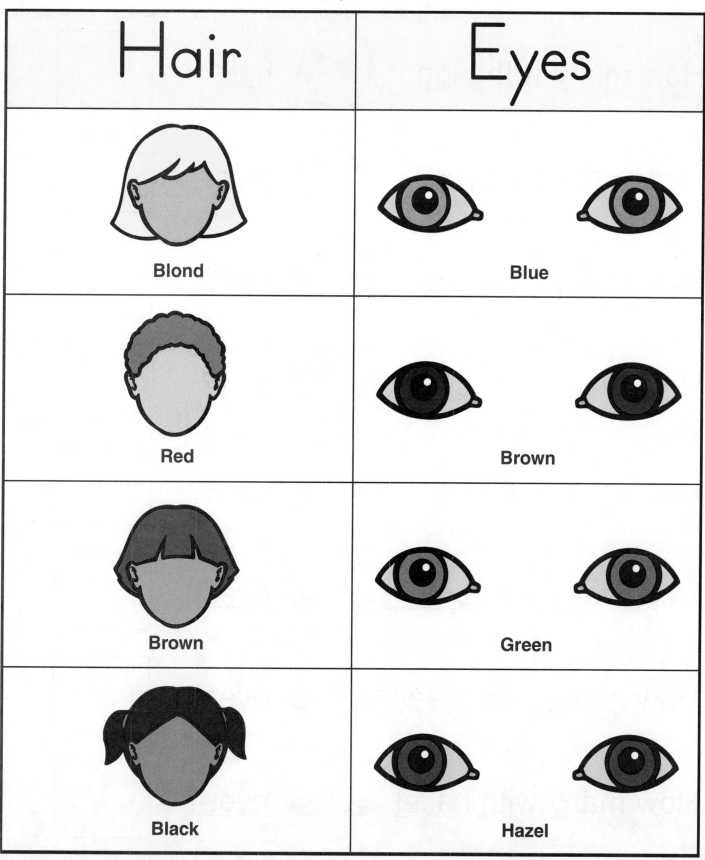

A Class Act

Directions: Look at the portraits.
Count the number with each attribute and write the number in the box.

How many with <u>blond</u>  <u>hair</u>?

How many with <u>red</u>  <u>hair</u>?

How many with <u>brown</u> <u>hair</u>?

How many with <u>black</u>  <u>hair</u>?

How many with <u>blue</u> <u>eyes</u>?

How many with <u>brown</u> <u>eyes</u>?

How many with <u>green</u> <u>eyes</u>?

How many with <u>hazel</u> <u>eyes</u>?

#3398 Literacy Centers for Math Skills

Vocabulary Cards

compare	attributes
different	count
sort	same

Vocabulary Cards

A Class Act

A Class Act

A Class Act

A Class Act

A Class Act

A Class Act

A Class Act

A Class Act

Dish It Out

Objective
Students will develop logical reasoning by identifying the number of possible outcomes of a problem.

Materials
- cup and bowl cards (pages 21 and 23)
- age-appropriate recording sheets (pages 18 and 20)
- crayons
- resealable plastic bags
- extension activity cards (page 25)
- vocabulary cards (page 27)

Arrangement
1. Cut out and laminate the cups and bowls.
2. Separate the cup and bowl cutouts and store them in different bags.
3. Copy one recording sheet for each student.
4. Make sure students have the appropriate crayons.

Presentation
Each student begins the activity with two different colored cup cutouts. He or she should *identify* the number of ways the two cups can be stacked (red on top of yellow, or yellow on top of red) by *counting* the solutions. Then on the appropriate recording sheet, students color and/or write the number of *outcomes*. Students will then be asked to *solve* the activity with the three different colored bowls. (There will be six outcomes.)

Extensions
1. Set out manipulatives that have different colors or shapes, and ask students to identify the number of arrangement combinations.
2. Cut out and laminate the spoons, forks, and knives on page 25. Have students see how many different ways they can line up the utensils.

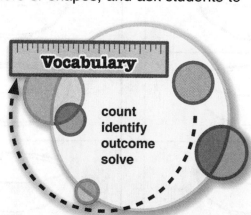

Vocabulary
- count
- identify
- outcome
- solve

©Teacher Created Materials, Inc.   17   #3398 Literacy Centers for Math Skills

Dish It Out

Directions: Color the ways you stacked the objects.

  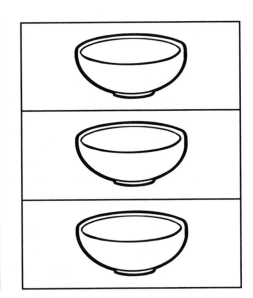

  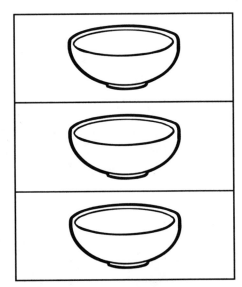

Dish It Out

1. Use the cards to stack the cups and bowls in different ways.

2. Answer the questions on the recording sheet.

©Teacher Created Materials, Inc.   19   #3398 Literacy Centers for Math Skills

Dish It Out

How many ways can two cups be stacked?

Draw the ways.

How many ways can three bowls be stacked?

Draw the ways.

Dish It Out

Dish It Out

Dish It Out   Dish It Out

Dish It Out   Dish It Out

Dish It Out   Dish It Out

Dish It Out

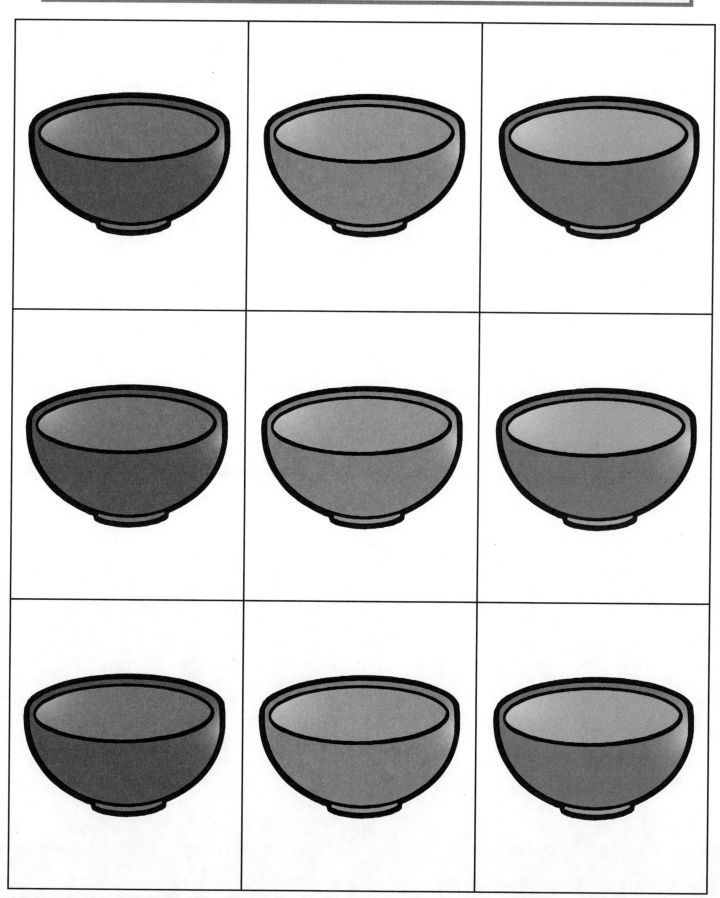

Dish It Out

Dish It Out   Dish It Out   Dish It Out

Dish It Out   Dish It Out   Dish It Out

Dish It Out   Dish It Out   Dish It Out

Dish It Out

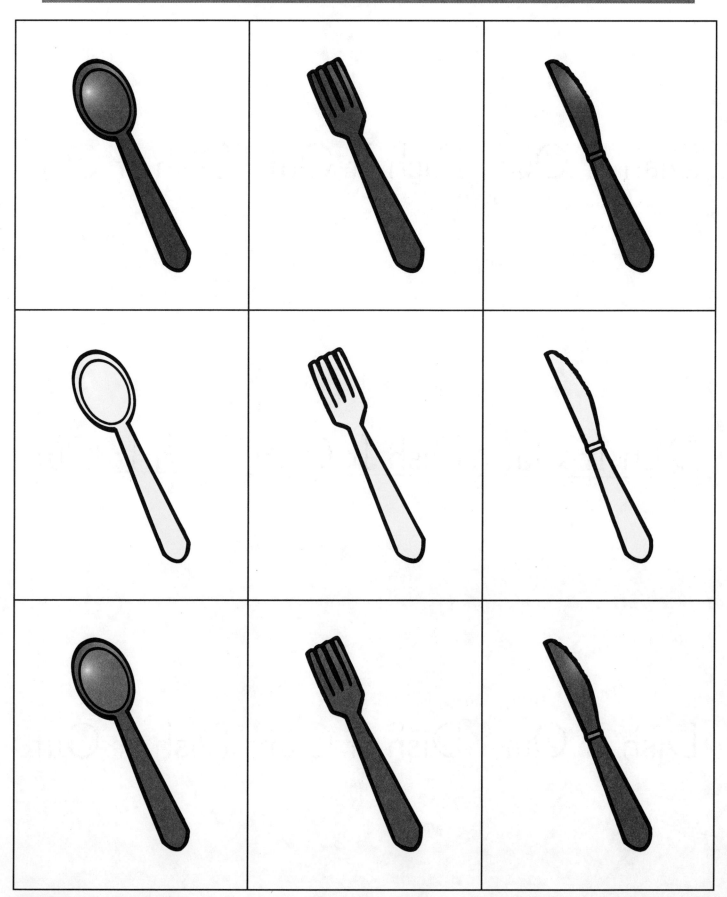

Dish It Out

Dish It Out   Dish It Out   Dish It Out

Dish It Out   Dish It Out   Dish It Out

Dish It Out   Dish It Out   Dish It Out

Vocabulary Cards

identify

solve

count

outcome

Vocabulary Cards

Dish It Out

Dish It Out

Dish It Out

Dish It Out

Barnyard Predictions

Objective
Students will learn to make predictions based on probabilities and record findings on a pictorial graph.

Materials

- age-appropriate recording sheets (pages 30 and 32)
- barnyard cards (page 33)
- container (cup or bag)
- extension sheet (page 36)
- vocabulary cards (page 37)

Arrangement

1. Make one copy of the recording sheet for each student.
2. Cut out and laminate the barnyard cards.
3. Place the cards in a container.
4. Cut out and laminate the color cube.
5. Tape or glue the cube together.
6. Make copies of the extension sheet.

Presentation

This center will help students understand how to predict outcomes based on the probability, or likelihood, of an outcome.

Ask the students to cut out the 15 barnyard cards. (Six cows, four horses, three pigs and two chickens are given.) Have the students *sort* the cards into piles by animal. Then ask students to put the cards in the container. They should *predict* which animal will be drawn the *most* number of times and which animal will be drawn the *fewest* times and record the predictions on the appropriate recording sheet. Each student draws eight cards, one at a time. After each card is drawn, the student *identifies* it and marks it on the given *graph*; then he or she returns the card to the container.

Extension

For more practice, cut out the color cube on page 35 and glue or tape it into a cube (3-dimensional) form. The paper cube has four sides that are red and two sides that are blue. Ask students to predict the color that will end up on top of the cube most often; then roll the cube 10 times, graphing the result on page 36 after each roll.

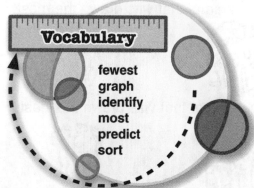

Vocabulary

fewest
graph
identify
most
predict
sort

©Teacher Created Materials, Inc.   29   #3398 Literacy Centers for Math Skills

Barnyard Predictions

Directions: Before you begin drawing cards, answer questions 1 and 2.

1. Circle the animal you suppose will be drawn from the container the most.

2. Circle the animal you think will be drawn from the container the fewest number of times.

3. Each time you draw an animal card, color it on the graph.

🐄	🐄	🐄	🐄	🐄	🐄
🐎	🐎	🐎	🐎		
🐖	🐖	🐖			
🐓	🐓				

4. Circle the animal that was drawn the most.

5. Color the animal that was drawn the least.

Barnyard Predictions

1. Answer questions 1 and 2 on the recording sheet.

2. Take out eight cards from the container, one at a time.

3. Color in one animal or block for the card drawn.

4. Finish a recording sheet.

©Teacher Created Materials, Inc.   31   #3398 Literacy Centers for Math Skills

Barnyard Predictions

Directions: Before you begin, answer questions 1 and 2.

1. Which animal do you think will be drawn from the container the most? _____

2. Which animal will be drawn from the container the fewest number of times? _____

3. Each time you draw an animal card, color a block on the graph.

Cow	**Chicken**	**Horse**	**Pig**

4. Which animal was actually drawn the most? _____

5. Which animal was actually drawn the least? _____

Barnyard Predictions

Barnyard Predictions

Barnyard Predictions	Barnyard Predictions	Barnyard Predictions
cow	cow	cow

Barnyard Predictions	Barnyard Predictions	Barnyard Predictions
cow	cow	cow

Barnyard Predictions	Barnyard Predictions	Barnyard Predictions
horse	horse	horse

Barnyard Predictions	Barnyard Predictions	Barnyard Predictions
pig	pig	horse

Barnyard Predictions	Barnyard Predictions	Barnyard Predictions
chicken	chicken	pig

Barnyard Predictions

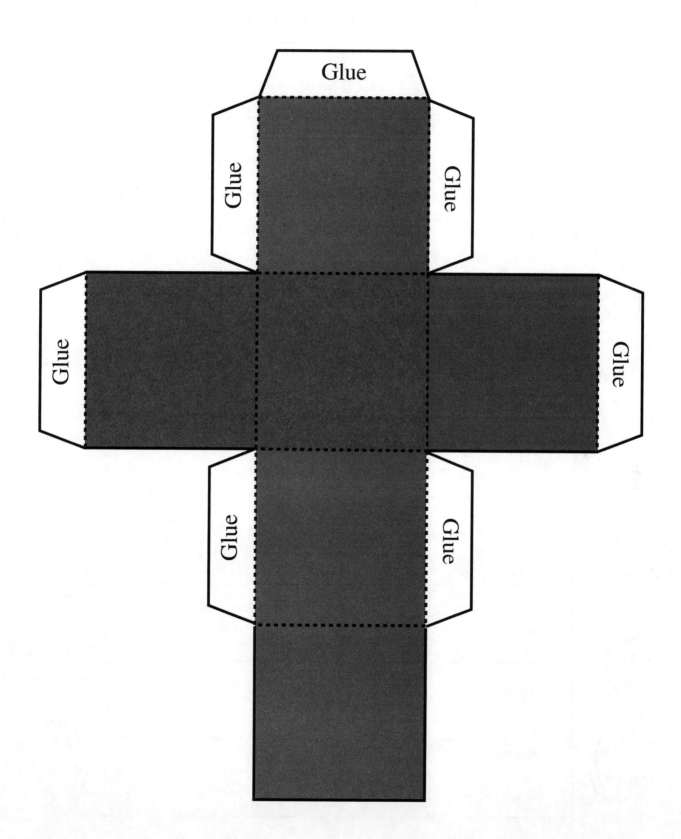

Barnyard Predictions

Directions: Take turns rolling the cube 10 times each.
Mark your results on the graph below after each roll.

Blue	Red

Vocabulary Cards

graph	fewest
most	identify
sort	predict

Vocabulary Cards

Barnyard
Predictions

Barnyard
Predictions

Barnyard
Predictions

Barnyard
Predictions

Barnyard
Predictions

Barnyard
Predictions

Barnyard
Predictions

Barnyard
Predictions

The Survey Says

Objective

Students will collect and use data for problem solving.

Materials

- tally chart (page 42)
- recording sheets (pages 43–50)
- crayons
- extension sheet (page 40)
- toothpicks
- dice
- vocabulary cards (page 51)

Arrangement

1. Make copies of the recording sheets for each student.
2. Laminate and cut out the tally chart.

Presentation

This center encourages the understanding of numbers, ways of representing numbers, relationships among numbers, and number systems. It also helps students create and use representations to *organize*, record, and communicate mathematical ideas.

Students will collect *data* and create *graphs* that will answer four questions. Ask the students to choose a topic and *survey* their classmates. Discuss the recording method (*tally marks*) that you would like them to use. Demonstrate how to record information using tally marks. Practice *counting* objects (five or more) to be certain students understand the different marks used. Demonstrate how to transfer the information collected to color in the bar graph for each survey. Model one survey and graph for the group before sending students to this center.

Extension

For more tally practice, allow students to roll a pair of dice and use toothpicks as tally marks to keep track of each roll on the extension sheet (page 40).

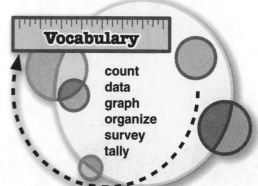

Vocabulary

count
data
graph
organize
survey
tally

The Survey Says

Directions: Roll the dice. Use the boxes for space to tally the numbers that are rolled.

1.

2.

3.

4.

5.

6.

The Survey Says

1. Ask classmates to answer the survey questions.
2. Tally the answers.
3. Color in the graphs with the survey findings.

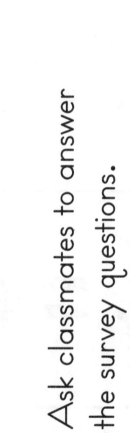
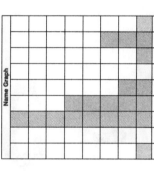
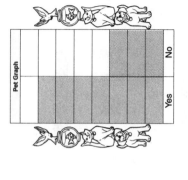

The Survey Says

I = 1	II = 2
III = 3	IIII = 4
✝✝✝✝ = 5	✝✝✝✝ ✝✝✝✝ = 10
✝✝✝✝ ✝✝✝✝ ✝✝✝✝ = 15	✝✝✝✝ ✝✝✝✝ ✝✝✝✝ ✝✝✝✝ = 20

The Survey Says

Food Survey

Directions: Ask each classmate the survey question. Record the answer by making a tally mark.

What is your favorite food?

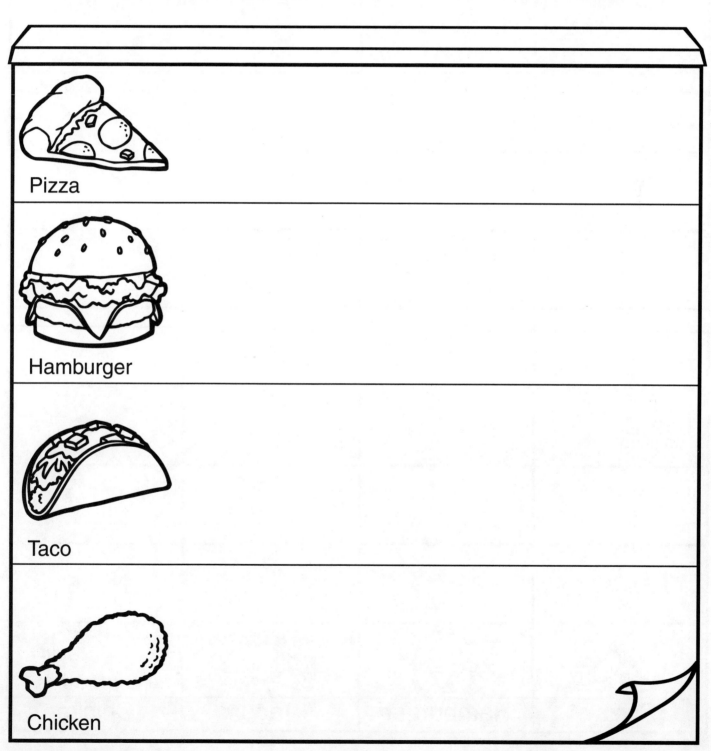

Pizza

Hamburger

Taco

Chicken

The Survey Says

Directions: Color in a block on the graph for each tally mark on the Food Survey.

Food Graph			
Pizza	Hamburger	Taco	Chicken

The Survey Says

Directions: Ask each classmate the survey question. Record the answer by making a tally mark.

Family Survey	
How many people are in your family?	
2	
3	
4	
5	
6	
7 or more	

©Teacher Created Materials, Inc.   45   #3398 Literacy Centers for Math Skills

The Survey Says

Directions: Color in a block on the graph for each tally mark.

Family Graph

2	3	4	5	6	7 or more

The Survey Says

Directions: Ask each classmate the survey question. Record the answer by making a tally mark.

	Name Survey
	How many letters are in your first name?
2	
3	
4	
5	
6	
7	
8	
9	
10 or more	

The Survey Says

Directions: Color in a block on the graph for each tally mark.

Name Graph

2	3	4	5	6	7	8	9	10 or more

The Survey Says

Directions: Ask each classmate the survey question. Record the answer by making a tally mark.

Pet Survey
Do you have a pet?
Yes
No

The Survey Says

Directions: Color in a block on the graph for each tally mark.

Pet Graph	
Yes	**No**

#3398 Literacy Centers for Math Skills ©Teacher Created Materials, Inc.

Vocabulary Cards

data	count
organize	graph
tally	survey

Vocabulary Cards

The Survey Says

The Survey Says

The Survey Says

The Survey Says

The Survey Says

The Survey Says

The Survey Says

Let's Go!

Objective

Students will estimate and measure the length of objects using non-standard units of measurement.

Materials

- vehicles cards (pages 57 and 59)
- connecting cubes
- appropriate recording sheets (pages 54, 61, and 62)
- vocabulary cards (page 63)
- crayons or markers
- optional: Measurement Strip (page 54)

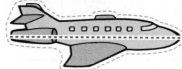

Arrangement

1. Laminate and cut out the vehicle cards.
2. Make a copy of the recording sheet for each student.
   Note: Delete the measurement strip (bottom of page 54) if manipulatives are available.
3. Arrange the connecting cubes, vehicle cards, and recording sheets in the center.
4. If connecting cubes are unavailable, have students create measurement strips using markers and the patterns on page 54.

Presentation

This center allows students working independently or in small groups to *measure* the lengths of different objects. (*Length* means "how long.") Explain to the students that each vehicle is a different length. Before the students begin measuring, ask them to look at the measurement tool they will be using and *estimate* (guess) the length of each vehicle.

(guess ⟶ *guesstimate* ⟶ estimate ⟶ measure)

Students will then use the connecting cubes (or the measurement strips) to measure the length of the red line on each vehicle. Demonstrate how to connect the cubes. Students may need to *round up* to the nearest cube, since connecting cubes vary in size.

Demonstrate how to record the measurement on the recording sheet. Color one box for each cube used. Then, fill in the amounts on the appropriate recording sheet and answer the questions.

Extensions

1. Find objects in the room that appear to be the same size as one of the vehicles. Once you have estimated, measure the object using the connecting cubes. Was your estimation correct?
2. Arrange the vehicle cards from shortest to longest.

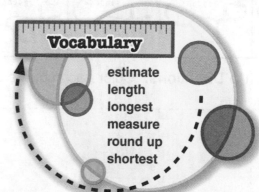

Vocabulary
- estimate
- length
- longest
- measure
- round up
- shortest

©Teacher Created Materials, Inc.   #3398 *Literacy Centers for Math Skills*

Let's Go!

Directions: Use the cubes or the Measurement Strip below to measure the length of each object. Use the dashed red lines as guides. Record your results by coloring one box for each cube used.

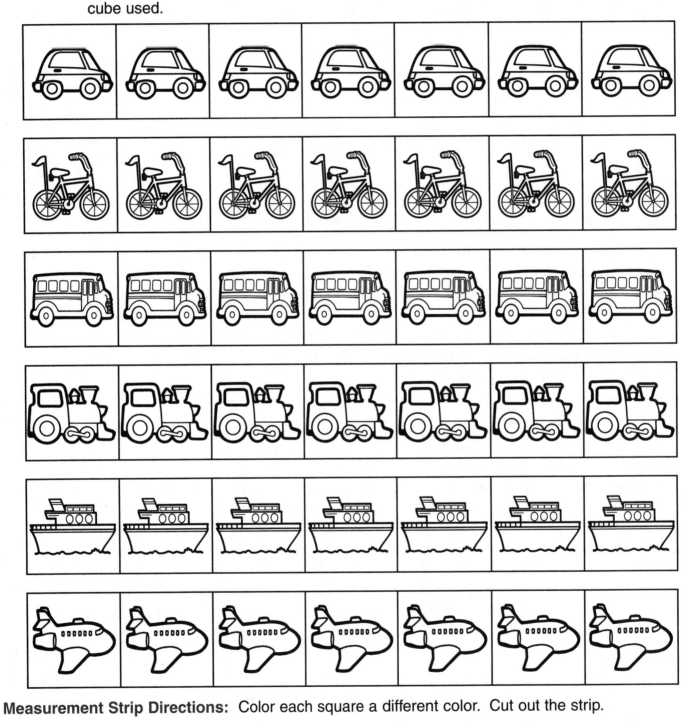

Measurement Strip Directions: Color each square a different color. Cut out the strip.

#3398 Literacy Centers for Math Skills © Teacher Created Materials, Inc.

Let's Go!

1. Measure the length of each object.

2. Record your results. Color one box for each cube used.

3. Answer the questions on the recording sheet.

Let's Go!

Let's Go!

Let's Go!

Let's Go!

Let's Go!

Let's Go!

Let's Go!

Let's Go!

Let's Go!

Let's Go!

Let's Go!

The car was  cubes long.

The bus  was cubes long.

The bike was  cubes long.

The train  was cubes long.

The boat  was cubes long.

The jet was  cubes long.

Let's Go!

1. Circle the longest vehicle.

2. Circle the shortest vehicle.

3. Which two vehicles were the same length?

4. Draw your favorite vehicle in the box below.

Vocabulary Cards

measure	estimate
longest	length
shortest	round up

Vocabulary Cards

Let's Go!

Let's Go!

Let's Go!

Let's Go!

Let's Go!

Let's Go!

Time for Fun

Objective

Students will identify time to the hour by using digital and analog clocks.

Materials

- analog clocks (pages 69 and 71)
- digital clock cards (pages 73 and 75)
- age-appropriate recording sheets (pages 66 and 68)
- whiteboard markers
- eraser or cloth
- extension sheets (pages 70 and 72)
- crayons
- vocabulary cards (page 77)
- resealable plastic bags

Arrangement

1. Cut out and laminate the two analog clocks.
2. Cut out and laminate the 24 digital clocks.
3. Copy one recording sheet for each student.

Presentation

This is an advanced lesson. Students will require prior knowledge about clocks and telling time.

Put students into partner groups. Two teams can work at the same time. Quickly review with students the difference between *analog* clocks, or those with *hands*, and *digital* clocks, those that numerically display the time. Students will take turns having 12 opportunities each to draw on the analog clock using the erasable pen. One partner will show the other a digital clock with a time. The partner will then draw the *hour* and *minute* hands on the clock to show the matching *time*. Then have students complete the appropriate recording sheet for analog clocks.

Extension

On the extension sheet, the lesson is reversed and students will be asked to find or write in the digital time to correspond with the analog clock shown on the sheet.

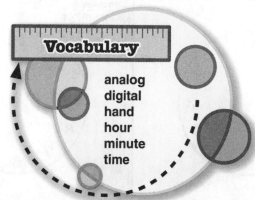

Vocabulary

analog
digital
hand
hour
minute
time

Time for Fun

Directions: Draw the hands on the analog clock to match the time.

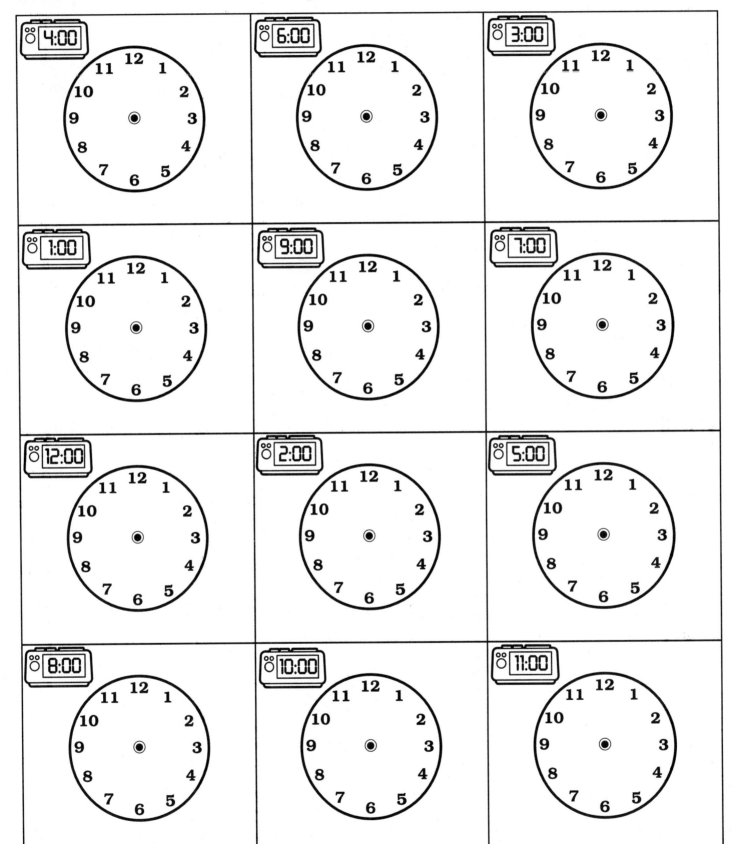

#3398 Literacy Centers for Math Skills 66 ©Teacher Created Materials, Inc.

Time for Fun

1. Show your partner a digital clock card.

2. Have him or her draw the hands on the analog clock to match the time.

3. Erase the clock.

4. Trade places with your partner and repeat.

Time for Fun

Directions: Color the analog clock that shows the correct time.

Time for Fun

Time for Fun

Directions: Color the digital clock that shows the correct time.

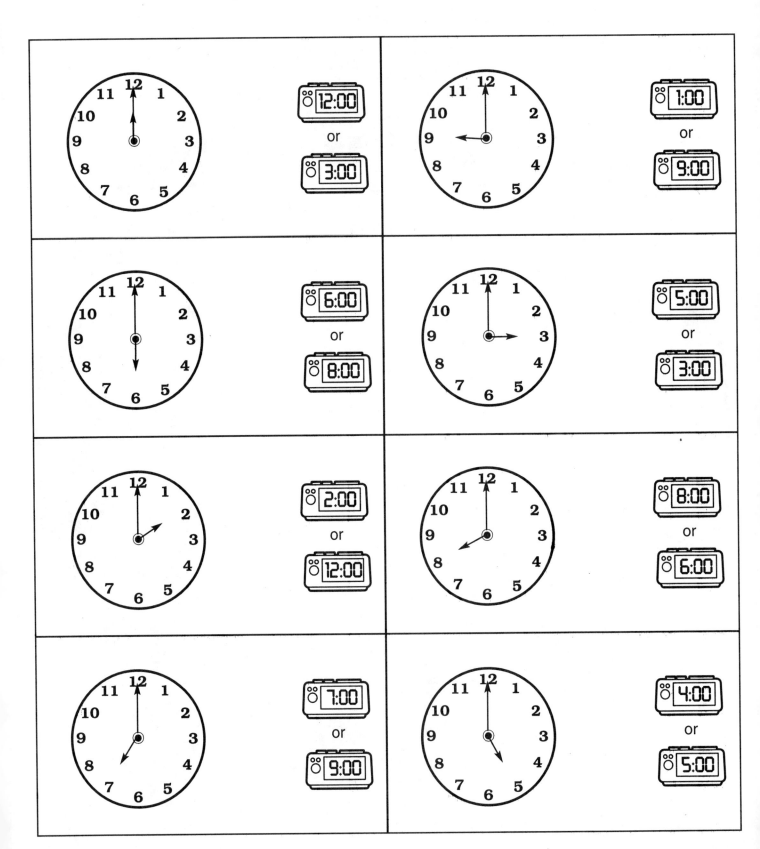

#3398 Literacy Centers for Math Skills

Time for Fun

Time for Fun

Directions: Write the matching time on each clock.

Time for Fun

Time for Fun

Time for Fun	Time for Fun	Time for Fun
Time for Fun	Time for Fun	Time for Fun
Time for Fun	Time for Fun	Time for Fun
Time for Fun	Time for Fun	Time for Fun

Time for Fun

Time for Fun

Time for Fun	Time for Fun	Time for Fun
Time for Fun	Time for Fun	Time for Fun
Time for Fun	Time for Fun	Time for Fun
Time for Fun	Time for Fun	Time for Fun

Vocabulary Cards

digital	hour	time
analog	hand	minute

Vocabulary Cards

Time for Fun

Time for Fun

Time for Fun

Time for Fun

Time for Fun

Time for Fun

Bugs!

Objective
Students will identify, create, and extend linear patterns.

Materials
- bug cards (page 83)
- resealable plastic bags
- bug pattern strips (pages 85–89)
- vocabulary cards (page 91)
- crayons or markers
- glue
- age-appropriate recording sheets (pages 80 and 82)

Arrangement
1. Laminate and cut out the bug cards and the bug pattern strips.
2. Make one copy of the recording sheet for each student.

Presentation

This center allows students to examine patterns and begin to analyze what relationships exist among them. The center will provide opportunities for students to use the information to predict what might happen next.

Explain to the students that a *pattern* is something that *repeats* itself (ababab; abcabc). Give a few examples for students. Consider directing their attention to a child who has a pattern on his or her clothing or arrange the students in a boy, girl, boy, girl pattern and see who can *identify* it. After a few examples, explain to the group that they will be identifying, creating, and extending patterns using the bug cards.

As practice, they are to finish the extended pattern on each of the bug pattern strips by using the bug cards. Explain that by looking at the existing pattern they can *predict* what will come next. When they have completed the patterns on the strips, they can use the cards to create their own patterns.

Extensions
1. Allow students to sort the bug cards and create their own patterns on the floor or workspace.
2. Set out stamps or stickers of bugs and allow students to create a variety of patterns on construction paper with them.

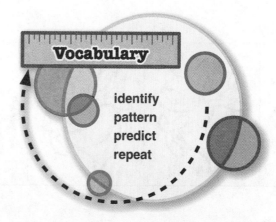

Vocabulary
identify
pattern
predict
repeat

©Teacher Created Materials, Inc.   #3398 *Literacy Centers for Math Skills*

Bugs!

Directions: Look at each pattern strip. Circle the bug that completes the pattern.

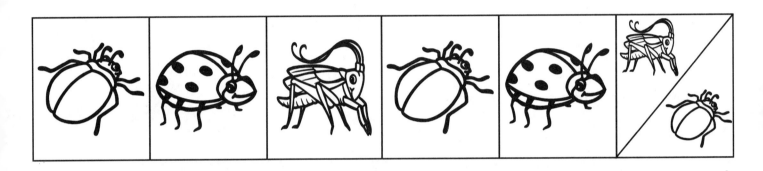

Bugs!

1. Sort the bug cards.

2. Use the cards to finish the patterns on the pattern strips.

3. Use the bug cards to make your own patterns.

Bugs!

Directions: Cut out the bug cards at the bottom.
Complete each pattern by gluing the correct bug at the end.

Bugs!

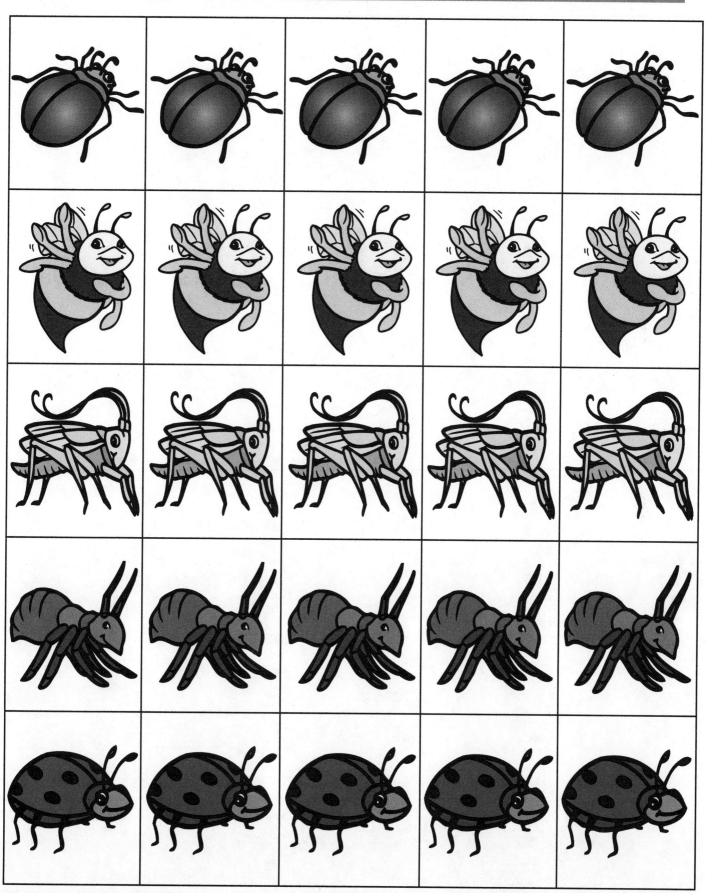

Bugs!

Bugs!	Bugs!	Bugs!	Bugs!	Bugs!
Bugs!	Bugs!	Bugs!	Bugs!	Bugs!
Bugs!	Bugs!	Bugs!	Bugs!	Bugs!
Bugs!	Bugs!	Bugs!	Bugs!	Bugs!
Bugs!	Bugs!	Bugs!	Bugs!	Bugs!

Bugs!

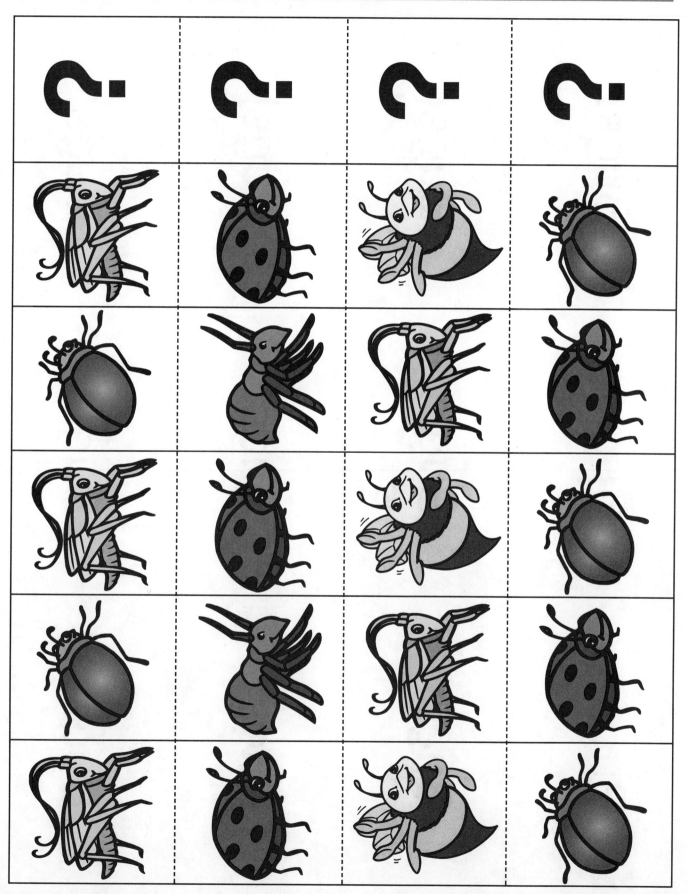

Bugs!

Bugs! Bugs! Bugs! Bugs!

Bugs! Bugs! Bugs! Bugs!

Bugs! Bugs! Bugs! Bugs!

Bugs!

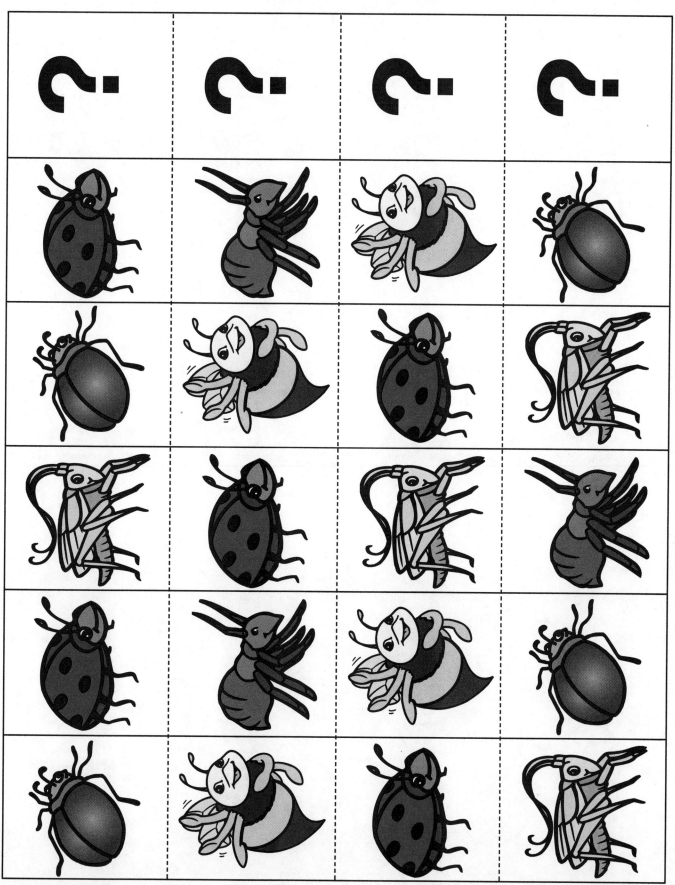

87 #3398 Literacy Centers for Math Skills

Bugs!

Bugs! Bugs! Bugs! Bugs!

Bugs! Bugs! Bugs! Bugs!

Bugs! Bugs! Bugs! Bugs!

Bugs!

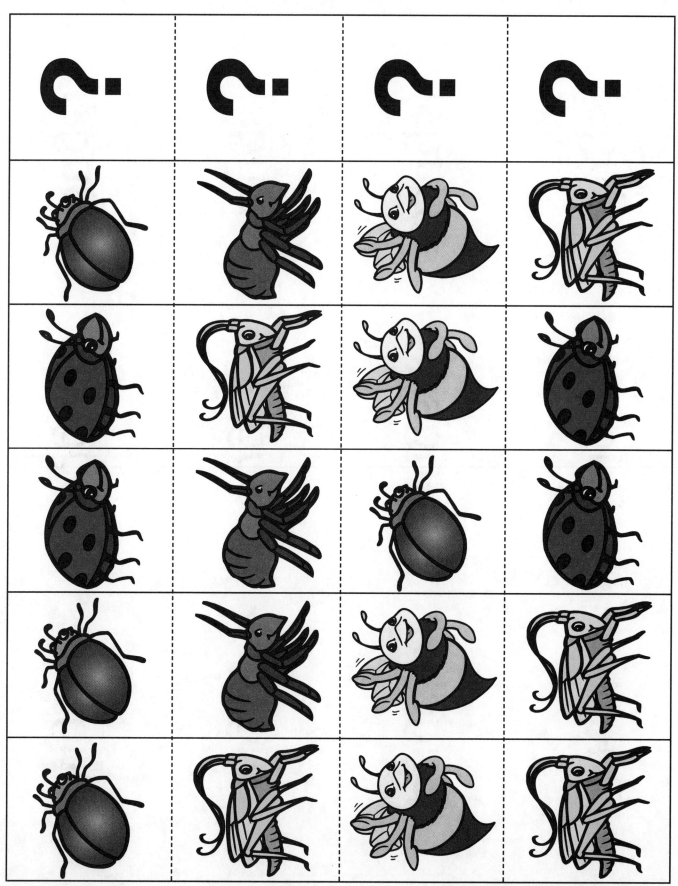

Bugs!

Bugs! Bugs! Bugs! Bugs!

Bugs! Bugs! Bugs! Bugs!

Bugs! Bugs! Bugs! Bugs!

Vocabulary Cards

pattern

repeat

identify

predict

Vocabulary Cards

Bugs!

Bugs!

Bugs!

Bugs!

Adding Shapes

Objectives
Students will develop problem solving strategies and learn how to do addition problems through joining sets.

Materials
- shape puzzles (pages 97–107)
- age-appropriate recording sheets (pages 94 and 96)
- extension sheets (pages 109–110)
- vocabulary cards (page 111)

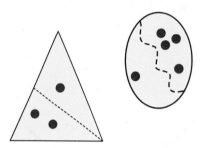

Arrangement
1. Cut out and laminate the shapes.
2. Cut the shapes into puzzle pieces following the dotted lines.
3. Copy one recording sheet for each student.

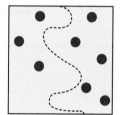

Presentation
Introduce the following shape vocabulary to your students before presenting the lesson. A *circle* is a round, closed shape without sides or corners. An *oval* looks like a face, and is longer or wider than a circle. A *rectangle* has four corners and four sides. Two sides are shorter than the others. A *rhombus* is a tilted shape with four equal sides. A *square* has four corners and four equal sides. A *triangle* has three sides and three corners.

Spread out the puzzle pieces on a table for the students. Ask the students to match up the puzzle pieces to create the correct shapes. Then students should add the "dots" inside the two puzzle pieces. On the recording sheet (pages 94 or 96), students should fill in the appropriate dots or numbers to correspond with the dots on the puzzle pieces (i.e., the top piece of the triangle has one dot, the bottom piece has two). Then, older students should finish the math equation revealed from the puzzle. In the case of the triangle, 1 + 2 = 3.

Extension
On the extension sheet (page 109), allow students to color each shape the appropriate color. On the second extension sheet (page 110), students will create addition problems using the number of shapes found on page 109.

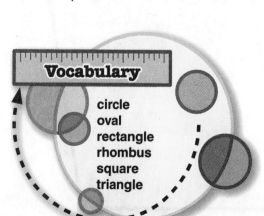

Adding Shapes

Directions: Draw the same number of dots on each shape as its puzzle.

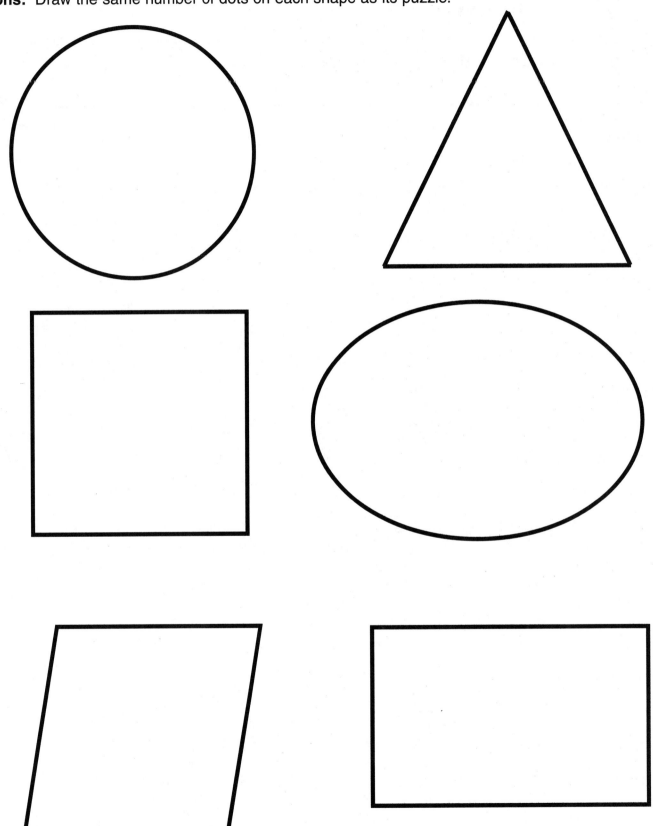

#3398 Literacy Centers for Math Skills   94   ©Teacher Created Materials, Inc.

Adding Shapes

1. Put each puzzle together to make a shape.

2. Count the dots on the two pieces.

3. Complete a recording sheet.

Adding Shapes

Directions: Fill in the dots for each puzzle piece.

Adding Shapes

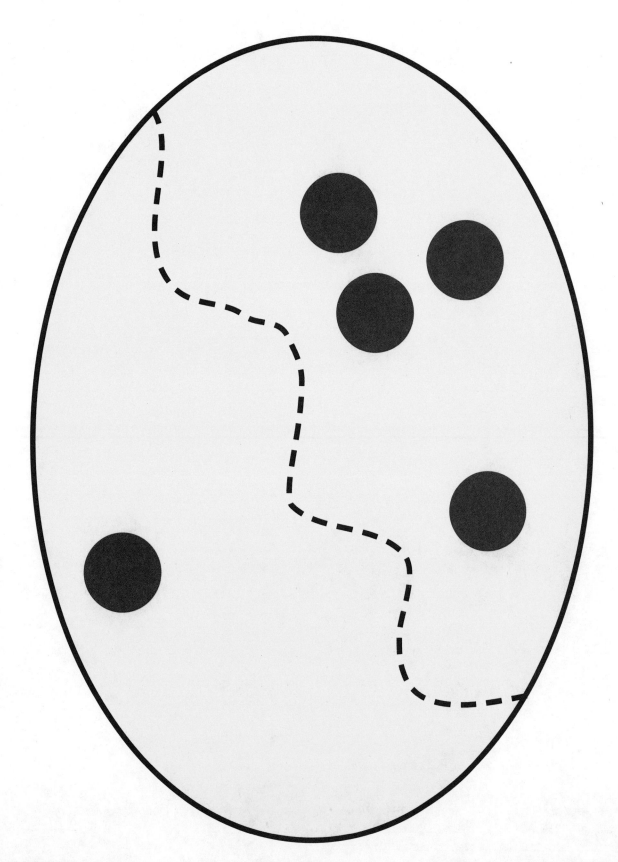

Adding Shapes

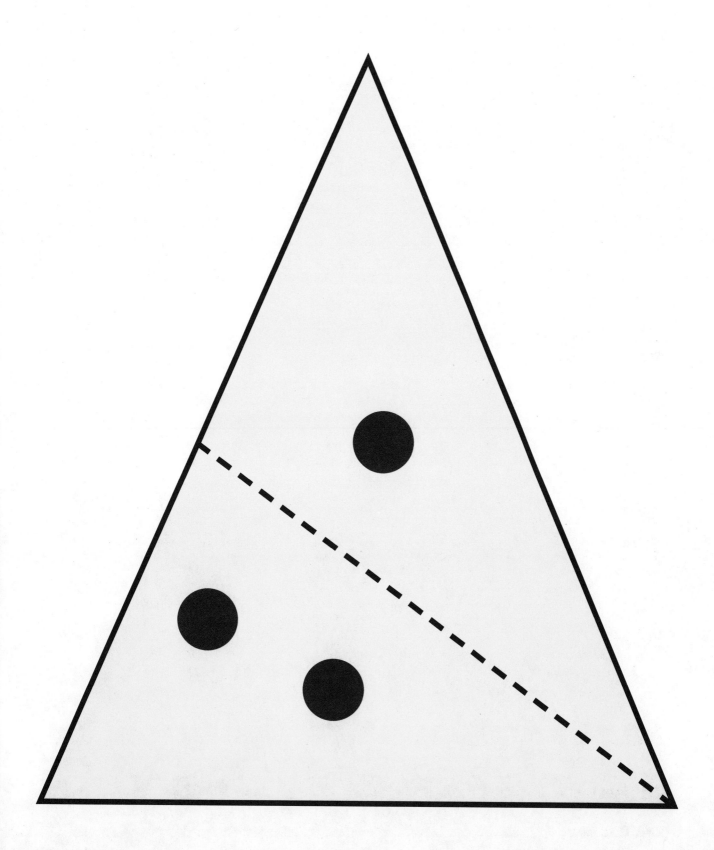

Adding Shapes

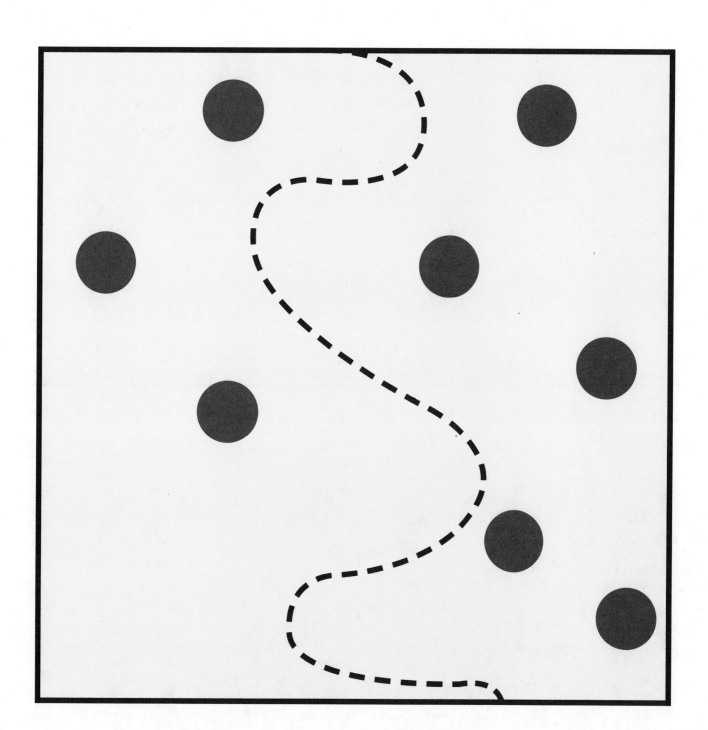

Adding Shapes

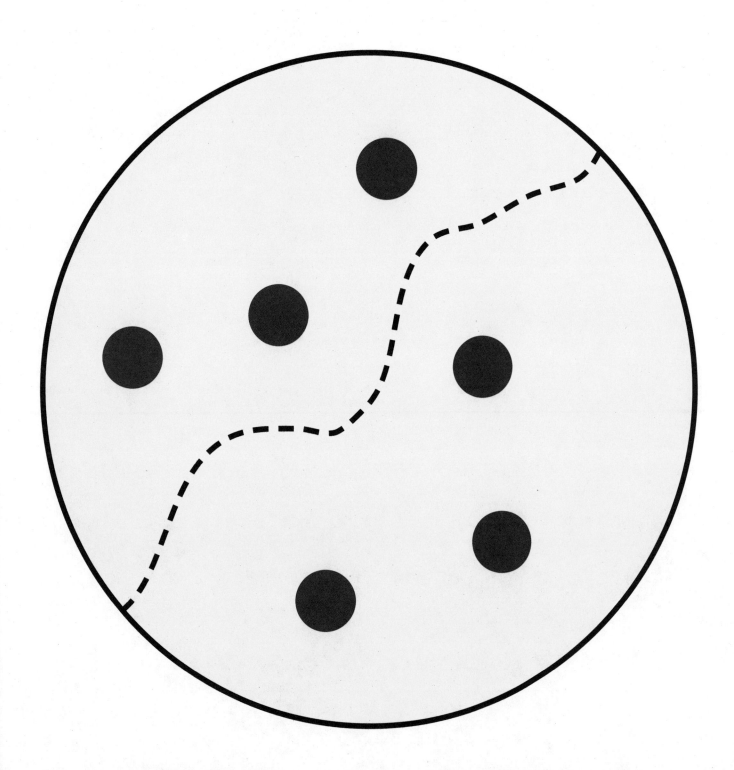

Adding Shapes

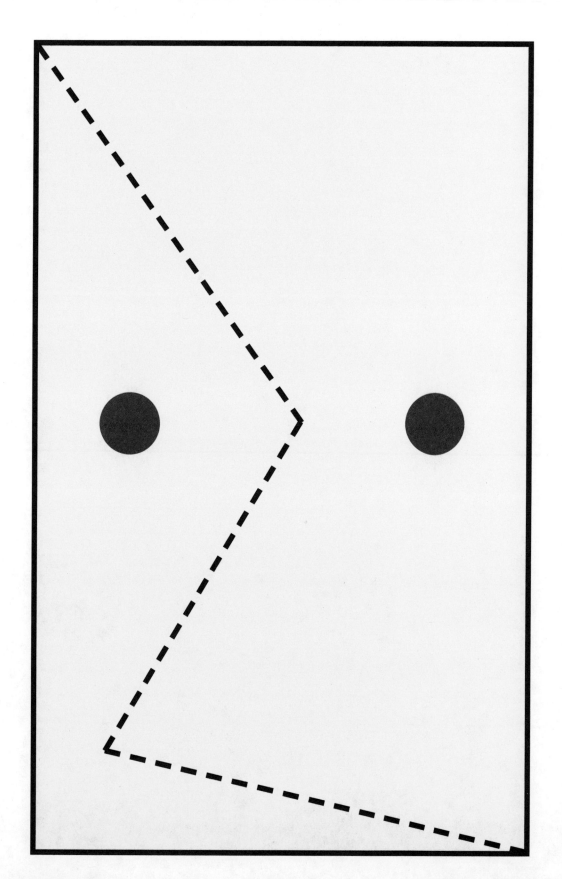

Adding Shapes

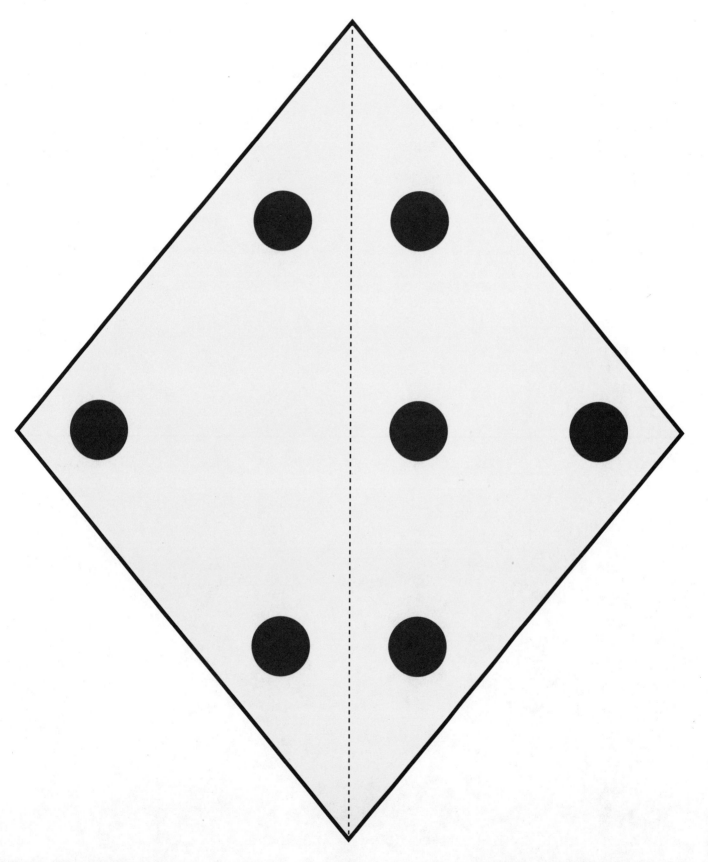

Adding Shapes

Color the circles ◯ blue.

Color the triangles △ orange.

Color the ovals ⬭ green.

Color the squares ☐ yellow.

Color the rectangles ▭ red.

Color the rhombuses purple.

©Teacher Created Materials, Inc.   #3398 Literacy Centers for Math Skills

Adding Shapes

Directions: Count the number of each shape on page 109 to create and solve these addition problems. Write the number inside the shape. Add the numbers.

Ovals + Triangles =

Rectangles + Circles =

Squares + Ovals =

Triangles + Circles =

Rectangles + Squares =

Rhombuses + Ovals =

#3398 Literacy Centers for Math Skills ©Teacher Created Materials, Inc.

Vocabulary Cards

circle	oval
rectangle	rhombus
square	triangle

©Teacher Created Materials, Inc.

Vocabulary Cards

Adding Shapes

Adding Shapes

Adding Shapes

Adding Shapes

Adding Shapes

Adding Shapes

Food for Thought

Objectives

Students will develop number sense by counting and comparing quantities to 12. Students will also learn to recognize numerals and number sets.

Materials

- number cards (pages 117–121)
- spinners (pages 123–125)
- 2 brads (1 for each spinner)
- age-appropriate recording sheets (pages 114 and 116)
- vocabulary cards (page 127)
- resealable plastic bags

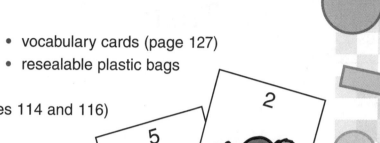

Arrangement

1. Cut out and laminate the number cards.
2. Divide the number cards evenly into two sets.
3. Cut out and assemble the spinners with the brads.
4. Copy one recording sheet for each student.

Presentation

This center will help students identify *greater than*, *less than*, and *equal to*.

The game is played between two students. Each player will start with 12 *numeral* cards (1 through 12). Players will take turns spinning the spinner and then turning over one card each. If the spinner is pointing to the "greater than" side, the player with the larger number card will add both cards to his or her pile. If the spinner is pointing to the "less than" side, the player with the lower number card will add the cards to his or her pile. If the students have the same card (equal), each will keep his or her own card and put it back in the pile. Students repeat until one student has collected all of the cards.

Following the game, the students again divide the cards so each student has a total of 12 cards (1 through 12). Individually, the students will turn over 2 cards at a time. Using the recording sheet on page 114, the student should write the smaller number in the "less than" column, and the higher number in the "greater than" column. Or, if working with younger students, have them complete the worksheet on page 116 by circling the correct answers.

Extension

For more practice to understand numerals and number sets, students can play the Go Shop game (similar to Go Fish). Each student starts with five cards. Students will take turns asking their partner for a set of cards. For example, "Do you have two bananas?" If the partner has that requested set, he or she must hand the cards over.

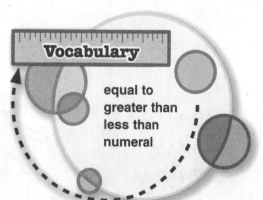

Vocabulary

equal to
greater than
less than
numeral

Food for Thought

1. Start with 12 cards numbered 1 through 12.
2. Turn over 2 cards at a time.
3. Write the smaller number in the Less Than column, and greater number in the Greater Than column.

Less Than	Greater Than

Food for Thought

1. Take 12 cards each.

2. Take turns spinning the spinner. Turn over one card each.

3. Look at the spinner to see who wins the cards.

4. Repeat until one player has all the cards.

Reminder:
If the spinner is pointing to "greater than," the player with the larger number will add both cards to his or her pile. If the spinner is pointing to "less than," the player with the lower number will add both cards to his or her pile. If the cards are equal, each person keeps his or her card.

Food for Thought

Directions: Circle the set in each row with the greater number of vegetables.

Directions: Circle the set in each row with the smaller number of fruits.

Food for Thought

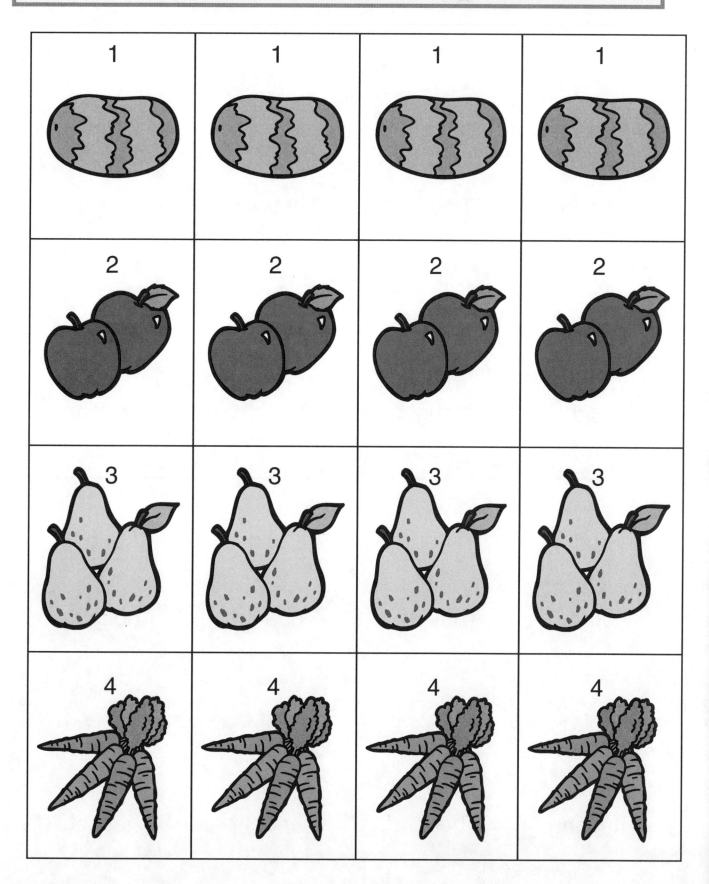

Food for Thought

Food for Thought	Food for Thought	Food for Thought	Food for Thought
Food for Thought	Food for Thought	Food for Thought	Food for Thought
Food for Thought	Food for Thought	Food for Thought	Food for Thought
Food for Thought	Food for Thought	Food for Thought	Food for Thought

Food for Thought

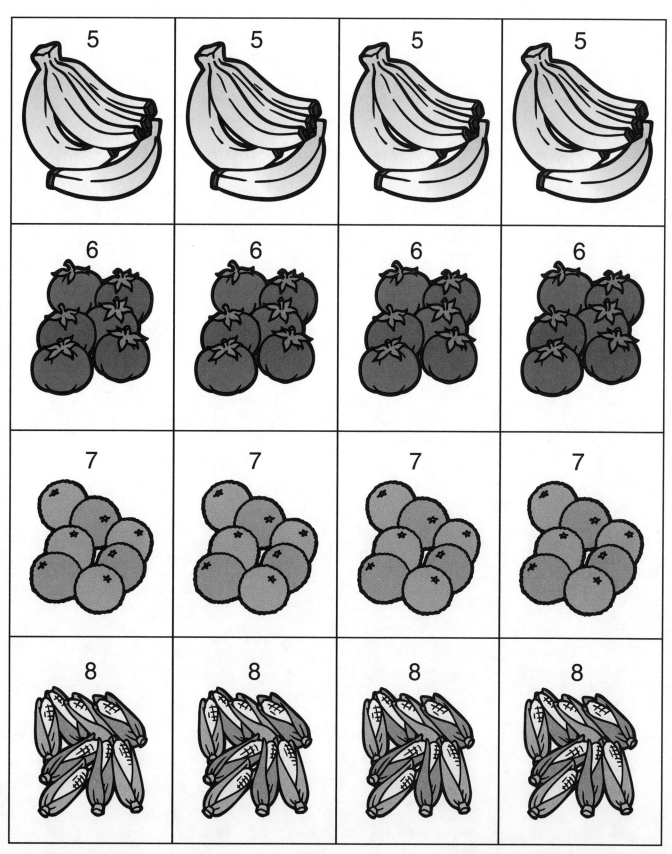

Food for Thought

| Food for Thought | Food for Thought | Food for Thought | Food for Thought |

| Food for Thought | Food for Thought | Food for Thought | Food for Thought |

| Food for Thought | Food for Thought | Food for Thought | Food for Thought |

| Food for Thought | Food for Thought | Food for Thought | Food for Thought |

Food for Thought

9	9	9	9
10	10	10	10
11	11	11	11
12	12	12	12

Food for Thought

Food	Food	Food	Food
for	for	for	for
Thought	Thought	Thought	Thought

Food	Food	Food	Food
for	for	for	for
Thought	Thought	Thought	Thought

Food	Food	Food	Food
for	for	for	for
Thought	Thought	Thought	Thought

Food	Food	Food	Food
for	for	for	for
Thought	Thought	Thought	Thought

Food for Thought

Directions: Cut out the circle and the pointer. Laminate both pieces. Use a brad to attach the pointer.

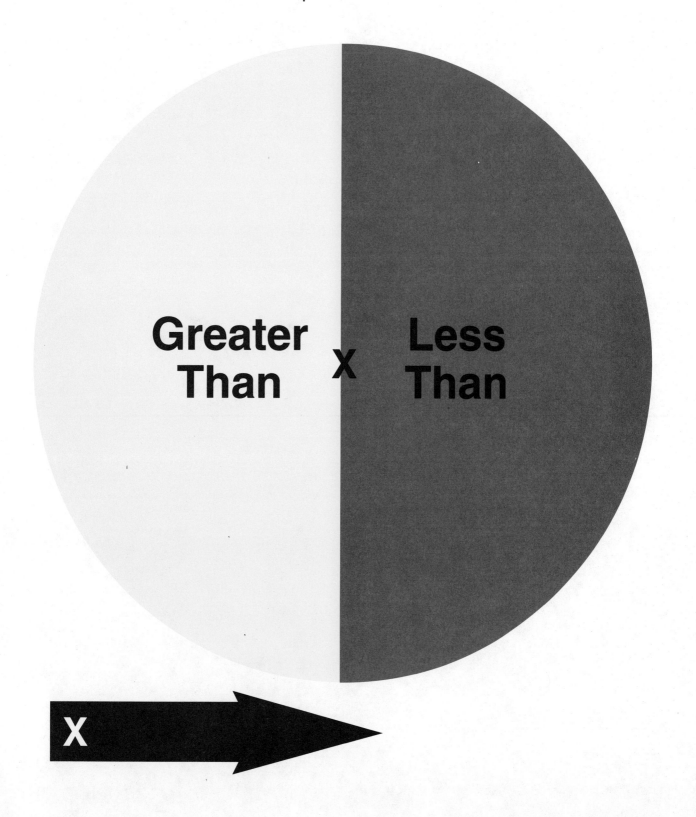

©Teacher Created Materials, Inc.   #3398 Literacy Centers for Math Skills

Food for Thought

Directions: Cut out the circle and the pointer. Laminate both pieces. Use a brad to attach the pointer.

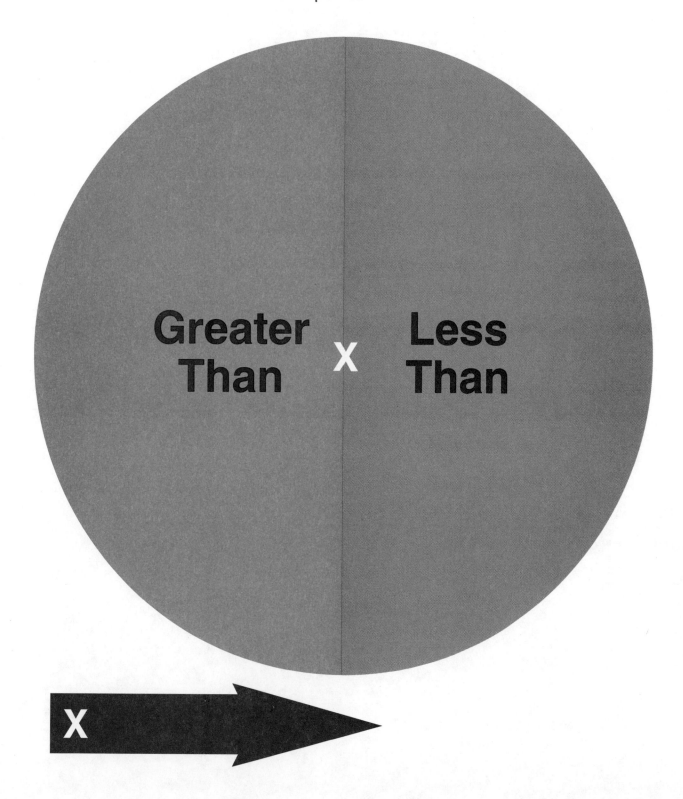

Vocabulary Cards

greater than

numeral

equal to

less than

Vocabulary Cards

Food for Thought

Food for Thought

Food for Thought

Food for Thought

Dino-Mite

Objective
Students will develop number sense by counting and comparing quantities and combinations of numbers up to 20.

Materials
- dinosaur game pieces (page 135)
- dinosaur game board (pages 137–139)
- Dino-Mite chips (page 133)
- age-appropriate recording sheets (pages 130 and 132)
- dice (number cubes)
- vocabulary cards (page 141)

Arrangement
1. Cut out, tape together, and laminate the game board.
2. Cut out, laminate, and assemble the dinosaur game pieces.
3. Cut out and laminate the Dino-Mite chips.
4. Set up the game board with the chips and game pieces before children arrive at the center.

Presentation
This center will help students *compare* quantities and number combinations to 20. Prior to introducing the game, allow students ample opportunities to practice rolling the dice (number cubes) and totalling the sums (dots).

Four students can play Dino-Mite at a given time. Students will take turns rolling the dice, *counting* the *set* of dots, and finding the sum. The student who rolls the dice can move any of the dinosaurs. The goal is to land a dinosaur *exactly* on the 20th box. This will take some practice. Choosing the right dinosaur to move may be a challenge at first. When a student succeeds in reaching the 20th box, he or she picks up a Dino-Mite chip and returns to the start. Play can continue until all the Dino-mite chips have been earned.

Extension
Allow students to practice number formation by rolling dice and writing down the sums. Each student finds the sum of his or her roll and compares it to the partner's sum.

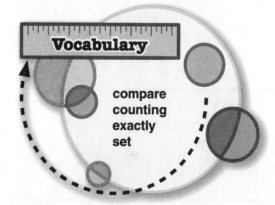

Vocabulary
compare
counting
exactly
set

©Teacher Created Materials, Inc.

Dino-Mite

Directions: After each roll, draw the dots on the dice. Find the sum.

| □ | + | □ | = _____ |

| □ | + | □ | = _____ |

| □ | + | □ | = _____ |

| □ | + | □ | = _____ |

| □ | + | □ | = _____ |

| □ | + | □ | = _____ |

| □ | + | □ | = _____ |

Dino-Mite

1. Take turns rolling the dice.
2. Find the sum and move a dinosaur piece.
3. Try to land on the 20th space.
4. Take a Dino-Mite chip when you land on 20. Then go back to the beginning.

Dino-Mite

Directions: Add the dice in each row. Write the sum.

1) 6 + 1 =

2) 4 + 2 =

3) 3 + 5 =

4) 1 + 1 =

5) 2 + 3 =

6) 2 + 2 =

Dino-Mite

Dino-Mite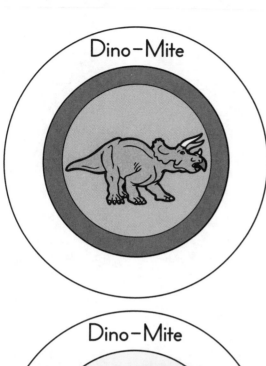

Dino-Mite

Dino-Mite

Dino-Mite

Dino-Mite

Dino-Mite

Dino-Mite

Dino-Mite   Dino-Mite

Dino-Mite   Dino-Mite

Dino-Mite   Dino-Mite

Dino-Mite

Directions: Cut out each circle and stand (strip). Cut the dashed lines on the stands. Tape the stands together to create rings. Slide the dinosaurs into the slits.

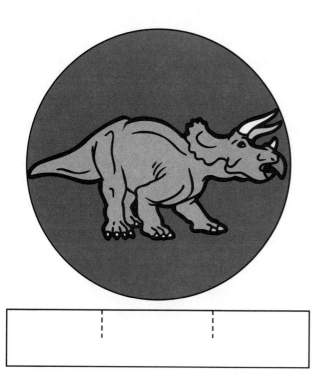

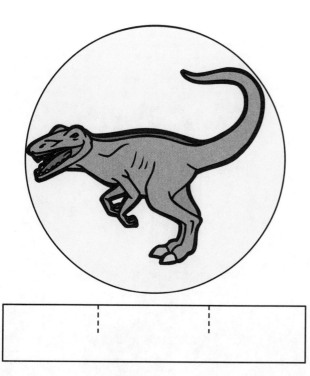

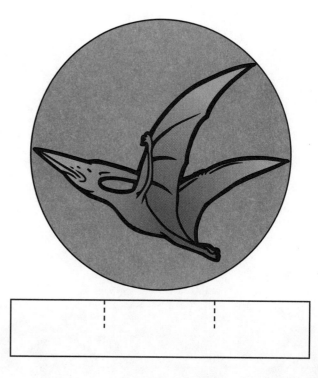

©Teacher Created Materials, Inc.   135   #3398 *Literacy Centers for Math Skills*

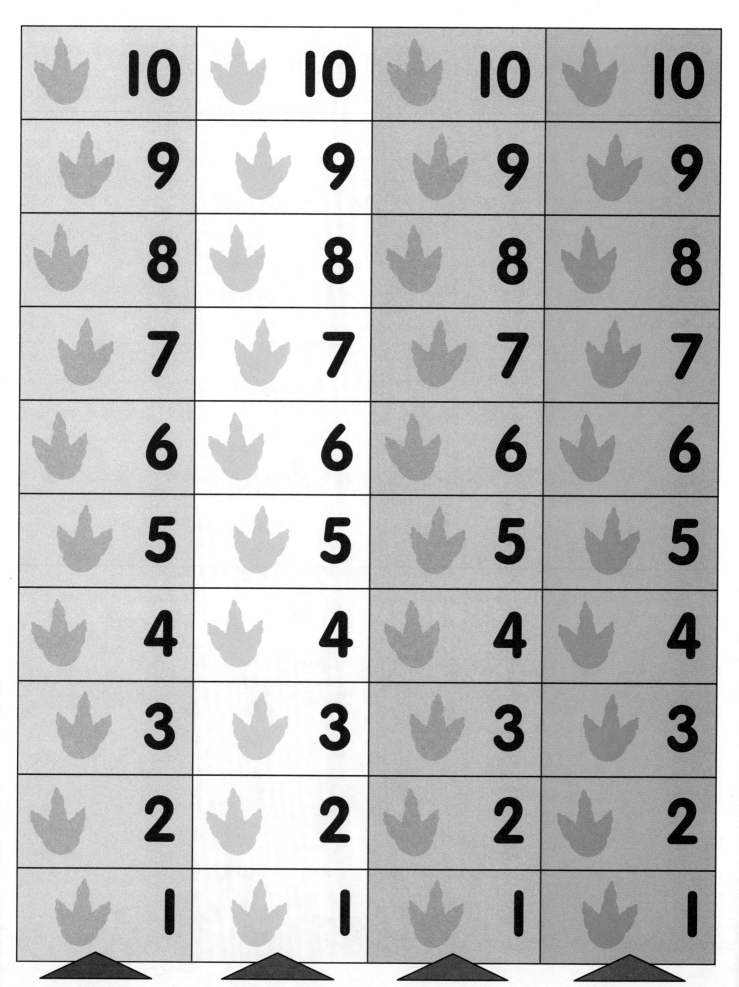

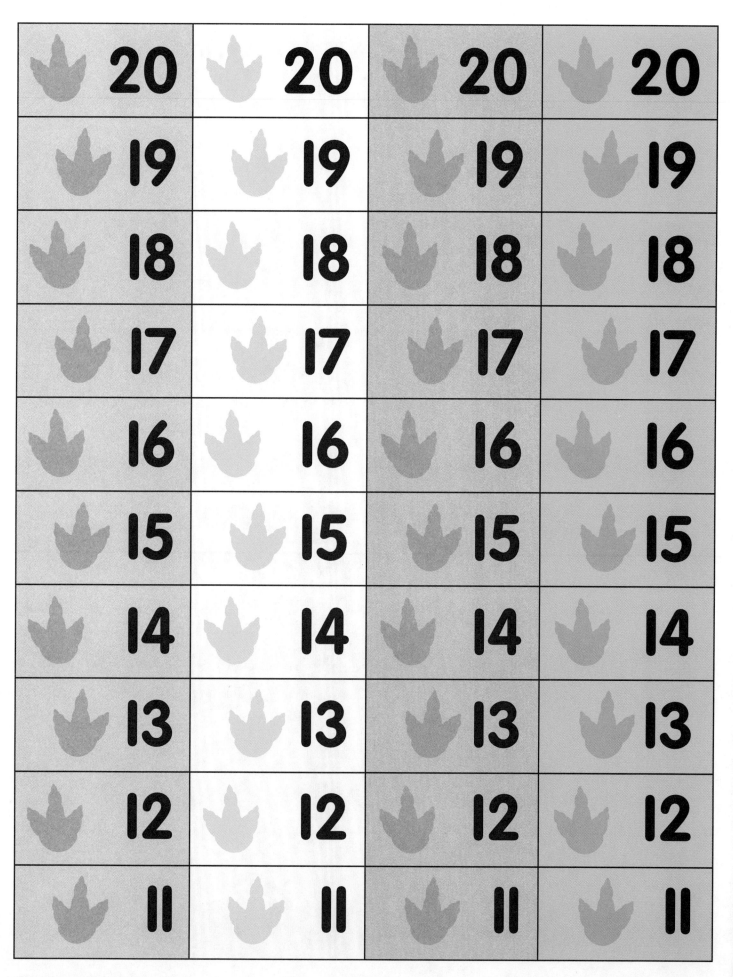

Vocabulary Cards

counting

set

compare

exactly

Vocabulary Cards

Dino-Mite

Dino-Mite

Dino-Mite

Dino-Mite

Hop to It

Objective
Students will use skip counting to identify missing numbers.

Materials
- carrot number cards (page 151)
- bunny number trails (pages 147–149)
- age-appropriate recording sheets (pages 144 and 146)
- vocabulary cards (page 153)
- resealable plastic bag

Arrangement
1. Laminate the bunny number trails.
2. Laminate the carrot number cards.
3. Copy one recording sheet for each student.

Presentation
This center encourages students to connect *patterning* to *skip counting*. Skip counting is a very important skill for young children to grasp. It involves creating patterns using a certain number. For instance, if one is skip counting by 2's the pattern would be 2, 4, 6, 8, 10, etc. One way to help students practice this skill would be to have them whisper the odd numbers and say the even numbers out loud. The same can be done skip counting with odd numbers and whispering the even numbers. Skip counting activities can be incorporated into reading the numbers on a number line, or the calendar. It can also be done when counting students lined up to go in or out of the classroom. Determine ahead of time if "odd" or "even" numbered students will call their number in whispered voices.

Students begin the activity with the bunny number trails and carrot number cards. Ask students to identify each missing number along the way by placing the appropriate card on the vacant spot. Ask students to trade cards when they are finished so each student at the center will have a chance to complete the bunny number trails. Then, have students complete the patterns on a recording sheet.

Extension
For more skip counting practice, allow students to use the carrot number cards to create their own patterns. One student lines up the cards in numerical *order* and removes cards to create a pattern. His or her partner tries to determine the numbers that complete the pattern.

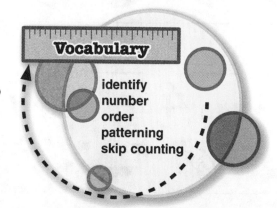

Vocabulary
identify
number
order
patterning
skip counting

Hop to It

Directions: Fill in the missing numbers in each row of the bunny's trail.

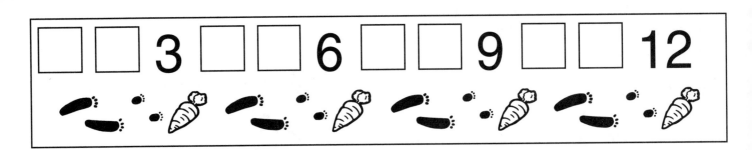

#3398 Literacy Centers for Math Skills ©Teacher Created Materials, Inc.

Hop to It

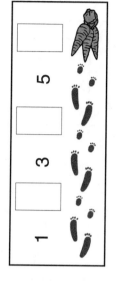

1. Use one of the four bunny number trails.

2. Put the correct carrot number card on each missing space.

3. When you have finished, trade bunny trails.

4. When you have done all four trails, fill out the recording sheet.

Hop to It

Directions: Circle the numbers that complete the pattern in each row.

 | 1 | 2 | 3/4 | 5/4 | 5 |

 | 2 | 4/5 | 6 | 8 | 10/12 |

 | 3 | 4 | 5 | 6/8 | 6/7 |

 | 5 | 9/6 | 7 | 8/11 | 9 |

 | 10 | 9/8 | 8/5 | 7 | 6 |

#3398 Literacy Centers for Math Skills ©Teacher Created Materials, Inc.

Hop to It

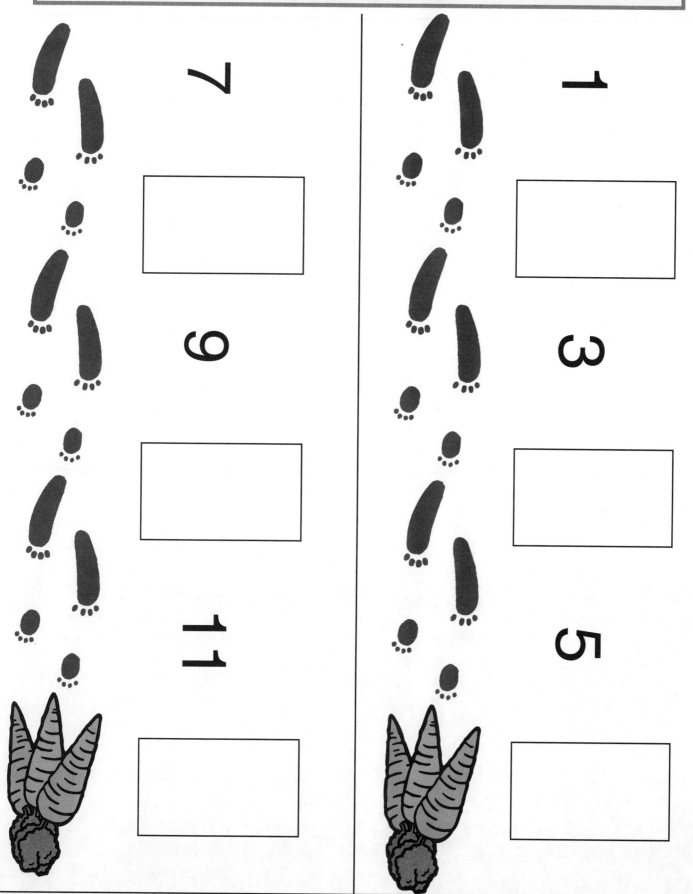

Hop to It

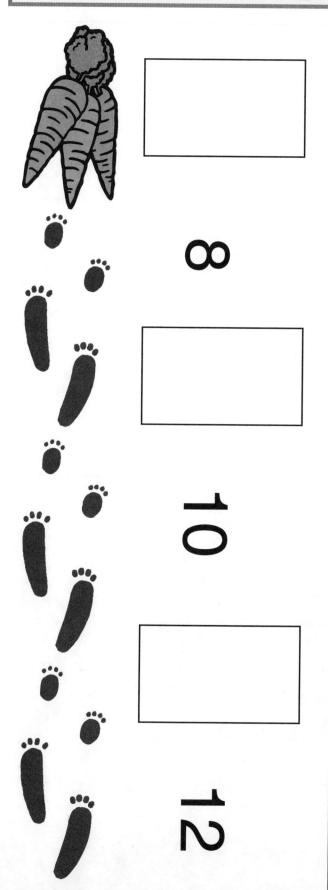
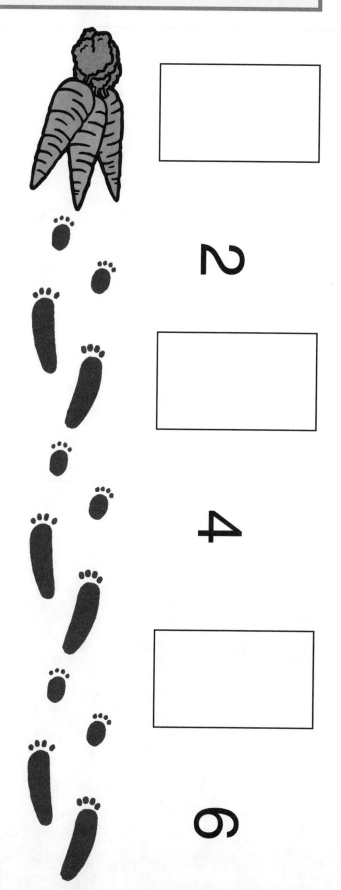

Hop to It

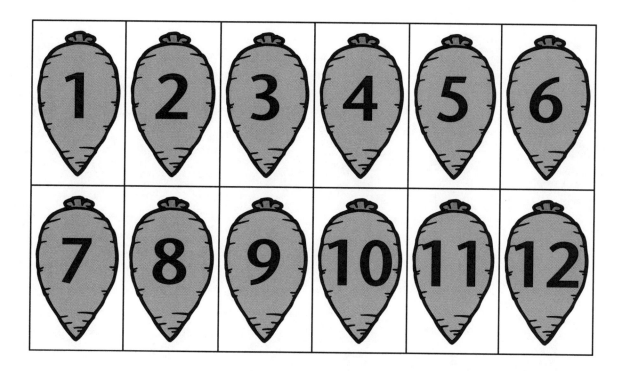

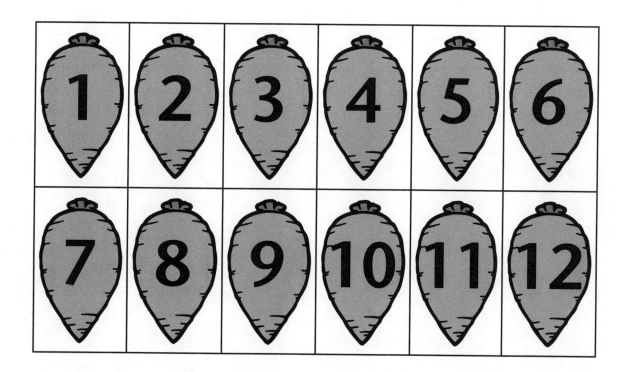

Vocabulary Cards

counting	identify
number	order
patterning	skip

Vocabulary Cards

Hop to It

Hop to It

Hop to It

Hop to It

Hop to It

Hop to It

Penguin Pals

Objective

Students will develop estimation skills by predicting how many objects will fit in a given space and comparing the estimate to the actual number.

Materials

- igloo patterns (pages 159–169)
- penguin patterns (pages 171–173)
- age-appropriate recording sheets (pages 156 and 158)
- resealable plastic bag
- cotton balls
- plastic cups (3 different sizes)
- vocabulary cards (page 175)

Arrangement

1. Laminate and cut out the igloo pieces.
2. Laminate and cut out the penguins.
3. Copy one recording sheet for each student.

Presentation

This center encourages the understanding of estimating *quantities*. At first, estimating is guessing. It is not unusual for students at this age to think that 100 people can fit in the school bus, even when the capacity is 25! The more hands-on estimating activities can be made, the better the estimates. Whenever possible, practice estimating with students. How many pencils can be lined up on the windowsill? How many lunch boxes will fit in a box? How many balls will fit in the storage bin?

On the appropriate recording sheet, students will *estimate* the *number* of penguins they believe will fit on the small, medium, and large igloos. Students will then use the penguin cards to find the *actual* number. Then ask students to record their findings on a recording sheet.

Note: Remind students that they can use penguins in different positions. They do not have to use one card style.

Extension

Allow students to do more exploring with estimation by using cotton balls (snowballs) and three different sized cups. Use the guess and check method. (Feel free to substitute cotton balls with any manipulatives available in your classroom.)

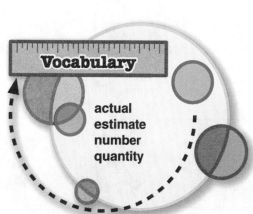

Vocabulary

actual
estimate
number
quantity

©Teacher Created Materials, Inc. — #3398 *Literacy Centers for Math Skills*

Penguin Pals

1. Estimate the total number of penguins that will fit on each igloo in the estimation box.
2. Place the penguins on the igloo.
3. Write the results, for the number (quantity) of penguins, in the actual number box.

Estimation

☐ penguins will fit on the **small** igloo.

☐ penguins will fit on the **medium** igloo.

☐ penguins will fit on the **large** igloo.

Actual Number

☐ penguins fit on the **small** igloo.

☐ penguins fit on the **medium** igloo.

☐ penguins fit on the **large** igloo.

#3398 Literacy Centers for Math Skills — ©Teacher Created Materials, Inc.

Penguin Pals

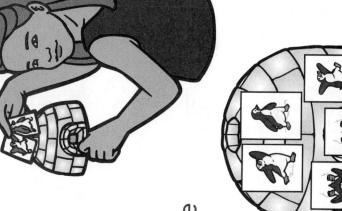

1. Estimate the number of penguins you think will fit on each igloo.

2. Write your predictions on the top half of the page.

3. Use the penguins to find the actual number that will fit on each igloo.

4. Record your findings in the second box.

Penguin Pals

1. Estimate the total number of penguins that will fit on each igloo in the estimation box.
2. Place the penguins on the igloo you have estimated.
3. Circle the results for the number of penguins in the actual number box.

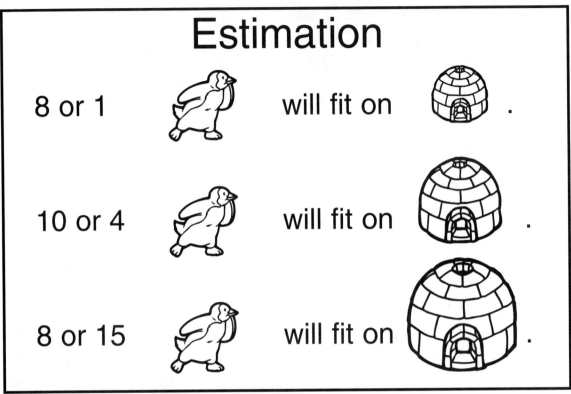

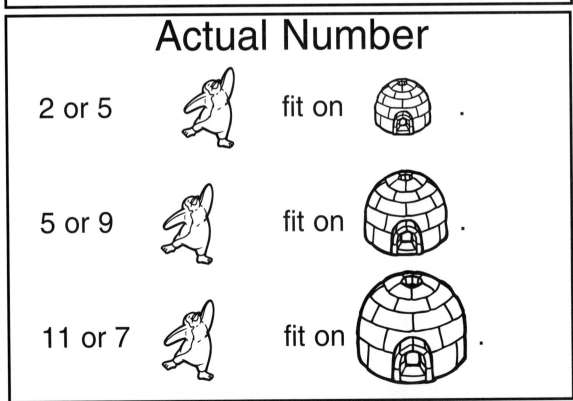

#3398 Literacy Centers for Math Skills

Penguin Pals

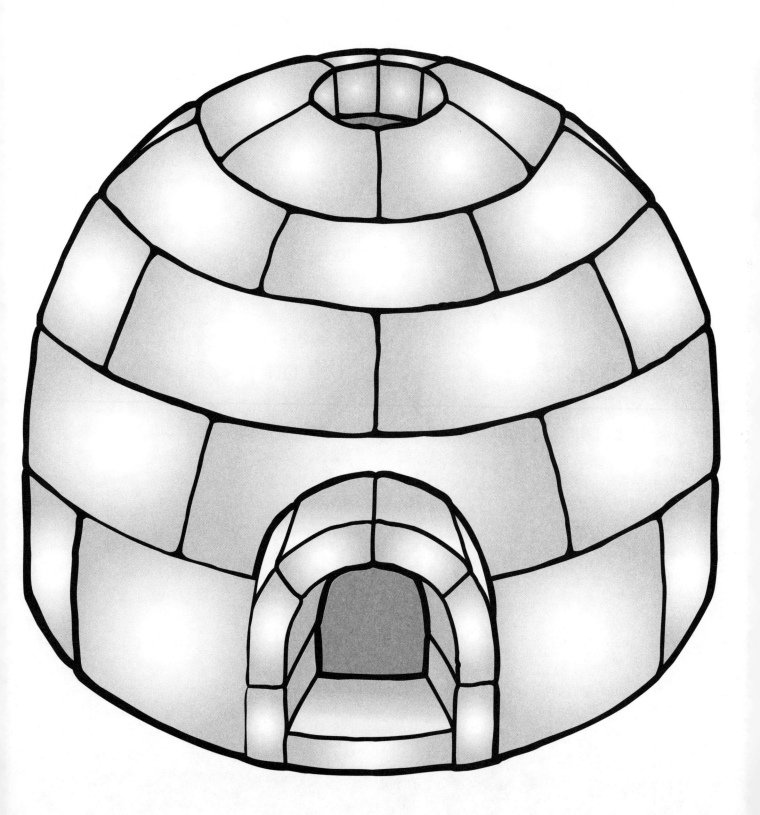

Penguin Pals

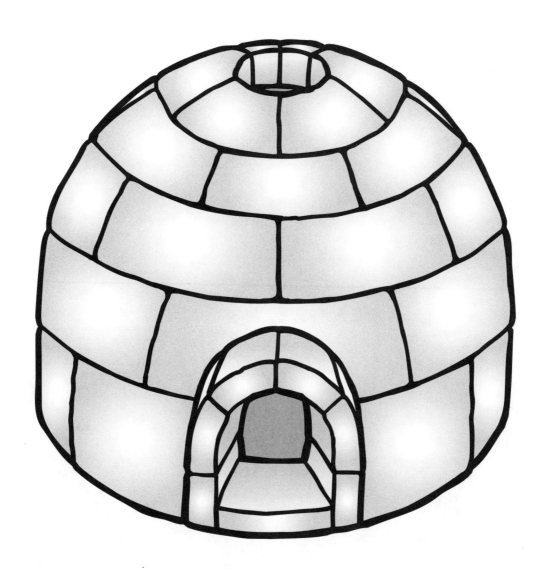

Penguin Pals

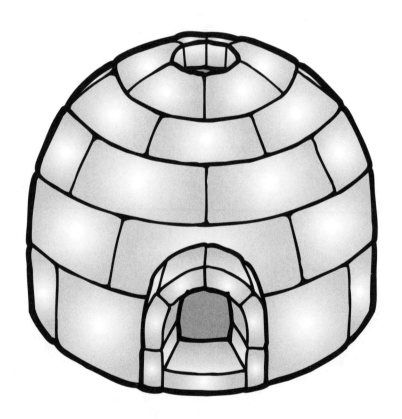

Penguin Pals

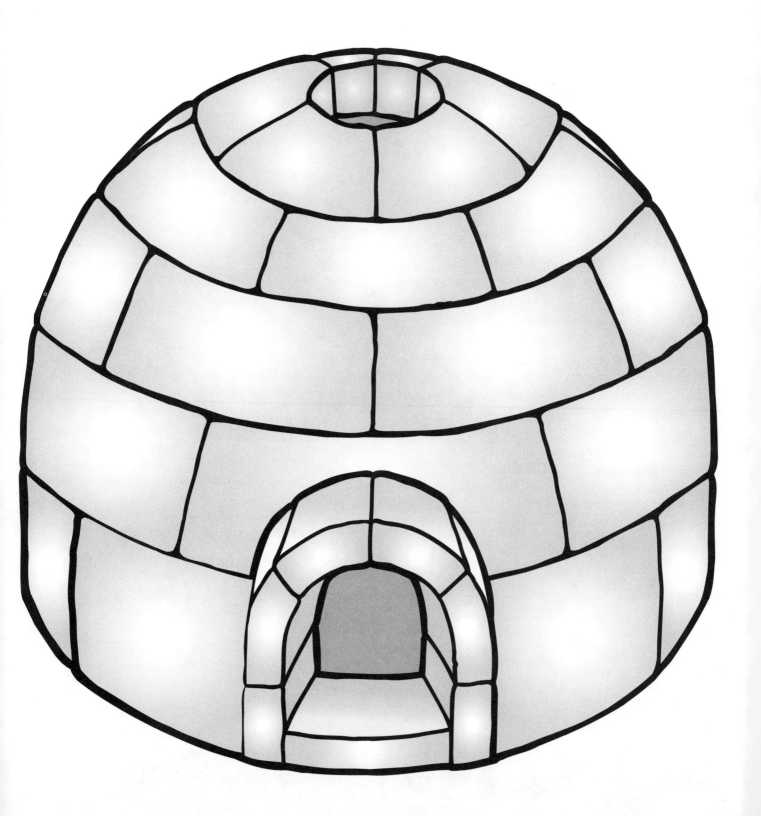

Penguin Pals

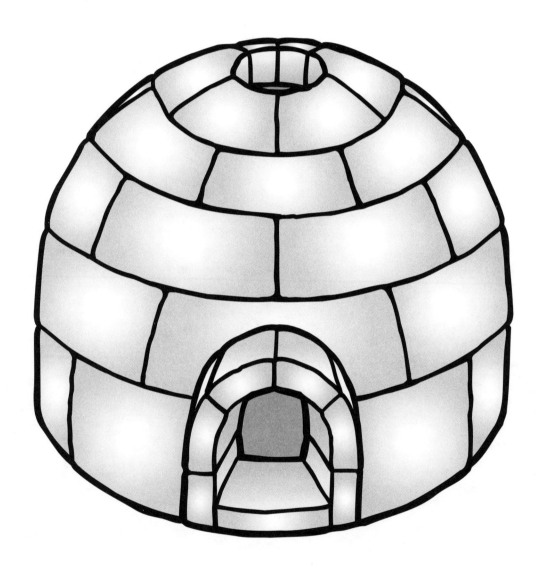

Penguin Pals

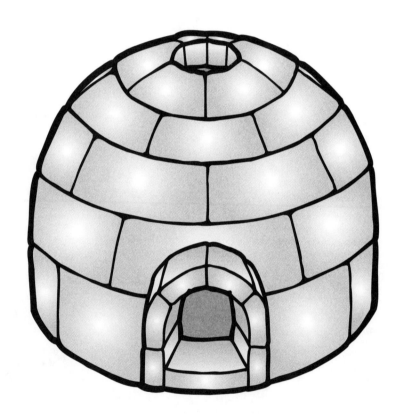

Penguin Pals

Penguin Pals

Penguin Pals	Penguin Pals	Penguin Pals	Penguin Pals
Penguin Pals	Penguin Pals	Penguin Pals	Penguin Pals
Penguin Pals	Penguin Pals	Penguin Pals	Penguin Pals
Penguin Pals	Penguin Pals	Penguin Pals	Penguin Pals

Penguin Pals

Penguin Pals

Penguin Pals	Penguin Pals	Penguin Pals	Penguin Pals
Penguin Pals	Penguin Pals	Penguin Pals	Penguin Pals
Penguin Pals	Penguin Pals	Penguin Pals	Penguin Pals
Penguin Pals	Penguin Pals	Penguin Pals	Penguin Pals

Vocabulary Cards

estimate	quantity
actual	number

Vocabulary Cards

Penguin Pals

Penguin Pals

Penguin Pals

Penguin Pals